THE STORY PARADOX

THE STORY PARADOX

HOW OUR LOVE OF STORYTELLING BUILDS SOCIETIES AND TEARS THEM DOWN

JONATHAN GOTTSCHALL

BASIC BOOKS
New York

Cover design by Chin-Yee Lai and Ann Kirchner
Cover image © VadimZosimov / Shutterstock.com

Basic Books
Hachette Book Group
1290 Avenue of the Americas, New York, NY 10104
www.basicbooks.com

Printed in the United States of America
First Edition: November 2021

Published by Basic Books, an imprint of Perseus Books, LLC, a subsidiary of Hachette Book Group, Inc. The Basic Books name and logo is a trademark of the Hachette Book Group.

The Hachette Speakers Bureau provides a wide range of authors for speaking events. To find out more, go to www.hachettespeakersbureau.com or call (866) 376-6591.

The publisher is not responsible for websites (or their content) that are not owned by the publisher.

Print book interior design by Linda Mark.

Library of Congress Cataloging-in-Publication Data
Names: Gottschall, Jonathan, author.
Title: The story paradox : how our love of storytelling builds societies and tears them down / Jonathan Gottschall.
Description: First edition. | New York : Basic Books, 2021. | Includes bibliographical references and index.
Identifiers: LCCN 2021010847 | ISBN 9781541645967 (hardcover) | ISBN 9781541645974 (ebook)
Subjects: LCSH: Persuasion (Rhetoric) | Storytelling—Social aspects. | Storytelling—Psychological aspects. | Rhetoric—Social aspects.
Classification: LCC P301.5.P47 G68 2021 | DDC 808—dc23
LC record available at https://lccn.loc.gov/2021010847

ISBNs: 9781541645967 (hardcover); 9781541645974 (ebook)

LSC-C

Printing 1, 2021

For Mom and Dad

Tell me a story.
In this century, and moment, of mania,
Tell me a story.

(Robert Penn Warren, "Audubon")

CONTENTS

INTRODUCTION

NOT TOO LONG AGO I WENT TO A BAR IN WHAT I SUSPECTED would be a doomed effort to simply think. I was feeling depressed about the state of the world and confused about this book. I'd been researching it and planning it for so long. I'd written many hundreds of pages of notes, and hundreds more of draft, and I'd tried and tossed a score of different titles as I named and rejected all the different versions of this book I might write. It was early in the *annus horribilis* of 2020 and I knew I was going to blow through my deadline, and probably the next one, too.

Of course, I knew generally what the book was about. Stories. All kinds of stories—fact, fiction, and the uncanny valley of narratives that cuts between. More specifically, it was about the dark power of stories to shape our minds in ways we can't always detect. But I'd taken on so much in my research—twenty-four hundred years of scholarship on Plato's *Republic*, the terrors of the Atlantic slave trade, weird midcentury panics about brainwashing, the hilariously terrifying rise of QAnon and flat earthism, the epidemic of mass shootings, deep dives into the artistic processes of some of the world's best (and worst) writers, the rise of virtual reality, the polarization of American society down clean narrative lines, along with reams of research on how our brains shape stories and are shaped by them.

This was all in pursuit of a question: Why, at this very moment, do stories seem to be driving our species mad?

So, I sat down at the quiet end of the bar and ordered a tall pour of the house's cheapest bourbon. I rolled in my earplugs and poised my pen over a clean bar napkin. I was waiting for the booze to kick in, hoping that altering my state of consciousness, along with my scenery, might bump me out of my creative rut. I stared at the napkin for a while. I doodled. And then I ordered another drink and looked around. First, I watched a chef-based reality show on one of the muted TVs. And then I turned to the next TV and watched two burly guys on ESPN shout at each other across a table. And then I turned a little farther to watch a cop show end and a news program begin.

By then, I'd admitted to myself that I was there to drink, not think. But as I gazed around the bar, I found that the booze really had given me the sideways view of the world I'd been hoping for. Normally, when we observe a crowd of people, we don't *really* observe the crowd. We zoom in on a sequence of individuals. Maybe a particularly pretty person captures our eye, and we study them for as long as we feel we can get away with it. And then our eyes flit to a person dressed especially stylishly or strangely. And then to an especially short or tall or thin or heavy person. Our eyes flick and flick again from one deviation from the norm to the next.

That's how it is for me anyway. But on this night, I was able to notice the crowd, not the individuals—the forest, not the trees. It makes human beings feel good to pretend that our behavior is various, diverse, unpredictable. But it's not. It's uniform, stereotyped, and predictable. And all the people in the bar (with the exception of your sad and soused author) were doing *exactly* the same thing.

Here's what they were doing. They were waving their hands in the air. They were opening and closing their mouths. They were moving their lips and tongues with great agility and stamina. Some of them were making trilling sounds. Others were bellowing. I saw a man cup a hand to a woman's ear and breathe warm information through the whorls directly to her brain. The woman's head snapped back in a kind of convulsion. Her neck veins popped as she woofed at the ceiling.

I was riveted. *Everyone* in the bar was doing this. The patrons at the tables. The servers and bartenders. Even the people on TV—the angry ex-jocks, the plasticized news readers, the actors playing cops and robbers, the guy pitching ShamWows.

I turned back to my bar napkin. I seated a squirming earplug with a finger poke. "These people are *weird*," I wrote. "Why are they here? What are they *actually* doing?"

Of course, I knew what they were doing. They were meeting their friends. They were trying to meet the loves of their lives. Or, like me, they were medicating their way through a bit of depression. But why is it that whenever people get together in groups as small as two, they are likely to fall into conniptions of lip flapping, face making, and hand waving?

Every day, all day, people move through gusts of words emitted by themselves and others. Humans spend their whole lives doing it—from the first sounds babies babble back and forth with their mothers to the last endearments croaked from our deathbeds. Whenever people get together, they take turns talking. And when we aren't ourselves conversing, we're mostly watching other people talk on TV or reading other people's words on a page like this one or listening to words spoken on a podcast or crooned in a song.

If you were an extraterrestrial scholar of human behavior and you were asked to name one activity that typifies humanity more than any other, you might say "they sleep" or "they work." But this would just show that you're an alien and you don't get us at all. If you were an Earth-born scholar of human behavior, like me, you might say "they communicate."

But, if you pay attention, most of these words seem unworth the breath. I was once invited to flap my hands and lips about the role of storytelling in effective teaching. The room I was flapping in was packed with safety instructors who happened to work at a nuclear power plant. *Their* words really matter. If the instructors don't choose precisely the right words, arranged in precisely the right order, a cataclysm could ensue.

Most of our communication isn't like this. Mostly, we're trading words about somebody's funny dog or dumb boss or frustrating boyfriend. The angry ex-jocks on the barroom TV were shouting about a hybrid football-folktale hero named Rob Gronkowski—should he stay retired or not? Even your sad and soused author was sitting at the bar babbling unstoppably with himself: Was he a big failure? How many books have killed their writers?

I stared down at the question I'd written on my napkin: **What are they actually doing?** I'd been tracing over the letters again and again until the question stood out in bold. I looked up and saw the people in the room standing up, sitting down, leaning forward or back, walking here and there. It was like they were being pushed by the breath of words, like a forest swaying in a breeze.

I wrote another word on my napkin: "Sway."

I turned the napkin slightly so I could look at the word from different angles, then added a question mark: "Sway?" I frowned down at the bar for a long time.

Sway

All the tireless communicating we do throughout our lives has a main overriding purpose. The purpose is to influence other minds—to sway how people think, feel, and ultimately behave. Whenever we communicate, we're using airy and insubstantial words to move each other around, if only a little, thereby rearranging the world in our favor.

We rarely have reason to breathe, type, or sing a word except sway. This applies even to the words we address only to ourselves. Although the inner voice is hard to study scientifically, psychologists long ago confirmed that "self-talk plays a role in inhibiting impulses, guiding courses of action, and monitoring goal progress."[1] In other words, self-talk is how we sway ourselves to shape up and fly straight.

In some cases, sway's place at the center of communication is obvious. When salespeople or politicians communicate, they're clearly trying to sway us toward a sale. But this is true, even if not so obvious, in settings like my bar. Although tavern talk is often microscopically small, it's still doing important work: it's establishing or maintaining the social alliances that have always been crucial to human flourishing. Even small talk with a stranger on a plane probably reflects an instinct to make a friend of a stranger before it occurs to the other person to kill us and steal our stuff.

We spend our whole lives blathering back and forth because our power as communicators is predictive of our sway—the degree to which we can move others to our will, instead of the other way around. Please realize that there's nothing necessarily, or even usually, Machiavellian about this. For all the rotten crap *Homo sapiens* gets up to, we're also, as the biologist Stephen Jay Gould puts it, "a remarkably genial species."[2] Our day-to-day interactions with friends and strangers are overwhelmingly kindly or at least neutral. One mark of this geniality is the determined efforts we make to sway other people for *their* good at least as much as ours. And humans, a once-meek species, inherited the Earth largely *because* sophisticated language allowed us to cooperate more ably than other animals.[3] In short, we seek sway mainly for symbiotic not parasitic purposes.

Maybe this talk of sway strikes you as interesting. Maybe it seems obvious—banal. Once the giddiness of my barroom breakthrough faded, the insight struck me as banal *and* interesting. But from this simple premise of the sway-making function of communication, it's a short step to a more radical syllogism about the purpose of storytelling.

Sway is the primary function of communication.

Storytelling is a form of communication.

Therefore, the primary function of storytelling is sway as well.

More than that, story is the main way that people—not just novelists or advertising mavens—seek to sway their fellows. After all, the people in the bar weren't exchanging bullet points or essays about the funny dog or the frustrating boyfriend—they were telling stories about them. The journalists on TV were telling stories, and the actors were acting them out, and the angry ex-jocks were working their Gronkowski points into the already rich soap opera of pro football in New England. And that reality show about the chefs wasn't really about cooking at all—it was about protagonists struggling toward happy endings, with the showrunners using every fair and foul trick to amp up the conflict and the drama.

The number one thing humans do is shoulder their way through winds of words. And these words aren't primarily organized into PowerPoint presentations, instruction manuals, spreadsheets, or lists. We spend huge chunks of the day, every day, moving through narratives—from a child's clumsy joke, to supermarket gossip, to marketing or propaganda campaigns, to the average four-plus hours Americans spend watching TV, to little chapters from our national and religious mythologies.

This is because stories are the single most potent way of influencing other minds. They're the best means humans have yet found for swaying one another so hard that we may stay bent forever. This is a wonderful thing when stories' huge capacity for good is channeled to promote empathy, understanding, charity, and peace. But the swayful magic of storytelling is just as good at sowing division, distrust, and hatred.

You are probably skeptical. And you should be. After all, I've claimed to be a scholar, but so far I've only told you a story, and if you leave this book with just one lesson, it should be this:

NEVER TRUST A STORYTELLER

But we do. And, really, we can hardly help ourselves.

Like puppies or rainbows, stories are one of the things that we all agree make life better. And this instinctive, unconditional infatuation is now being reinforced by a pancultural movement celebrating the transformative power of storytelling in business, education, law, medicine, self-improvement, and many other domains of human experience. But this wired-in lack of suspicion endows stories with a power that's stronger than rational argument and more irresistible than hard facts. When people are asked if stories influence how they think and act, most say, "Not so much."[4] Ironically, our bozo confidence that storytellers don't sway us is exactly what gives them such sway—sometimes for the better, but often for the worse.

I've already written a book, *The Storytelling Animal*, focused mainly on how stories sway us for the better. *The Story Paradox* revisits key topics from the earlier book but with a focus on how stories sway us for the worse, not just as individuals but especially at the level of whole societies.

So, what are we supposed to do about it? I'll address this question at the end. For now, I'll just name one fantasy this book won't entertain: Plato's famous notion that it would be possible to banish storytellers from society, or that we'd want to live in a world where we could. Human beings can no more give up narrative than we can breathing or sleeping. So throughout this book I'll not resist illustrating the power of stories in the most sway-making way I can. By telling you stories.

LARP

The man is forty-six years old. He's burly and puffy and utterly friendless. He's been unemployed for a while, but he wasn't always. And he wasn't always friendless, either. Once, the man worked at a bakery. His coworkers liked him. He went to their parties and played with their kids. His friends nicknamed him Babs. He'd never once been in trouble. Never hurt anyone. When he bought some very heavy guns, the background check raised no flags.

By October 27, 2018, the man wasn't as likable as he used to be. In fact, he was full of dark claims and urges. On that Saturday morning in October, I imagine him waking to a bachelor's breakfast of instant oatmeal or cornflakes. I imagine him sipping his coffee as he thumbs through his social media feeds, smiling at some posts, shaking his head at others. And then, after posting a message of his own—one to sway the whole world—he threw a bag in his truck and made the fifteen-minute drive to the leafy and prosperous Pittsburgh neighborhood of Squirrel Hill.

I wonder what he was thinking as he drove. Was he frightened of dying? Of killing? Did he pull over to think it through one last time? Was he seething with righteous certainty? Or was he just zoning out, like any other driver, as the traffic bottlenecked at the Fort Pitt tunnel?

The man parked in front of the Tree of Life Synagogue, made his way through the front doors, and started shooting. "All Jews must die!" he screamed, as he opened up with his AR-15 and three Glock .357 pistols.

In the few minutes it took for the first police to arrive, he killed eleven people and wounded many more. Shot by SWAT officers, the man would eventually surrender. As the officers worked to save his life, the killer tried to explain himself. He needed the cops to see that he wasn't one of these nihilistic mad dogs who blast their way through schools or shopping malls for no reason at all. If he'd committed a crime, he'd only done so to stop a far larger and more heinous one. The Jews were facilitating a de facto invasion of America and a slow-motion genocide against the white race.

That night, I made the half-hour drive to Squirrel Hill to take part in the hastily arranged vigil. I moved through thousands of mourners in a daze. *All of this*, I thought to myself, *this whole pall of death and sorrow is because of a story*.

The story is one of the world's favorite old goodies. It staggers through history like an undead thing—staggers forward no matter how hard we try to kill it, no matter how much evidence is brought forth to dispel it.

The story begins in the New Testament and goes forward in screed after screed—driving pogrom after pogrom—from Pontius Pilate washing his hands, to *The Protocols of the Elders of Zion*, to the manifestos of Stormfront.org. It's always the same story told in countless variants.

It goes like this: Jews are the vampires of history. With their satanic cleverness and covetousness, these subhuman superhumans are orchestrating a multimillennia conspiracy to enrich and empower themselves, and immiserate and enslave everyone else. If good people could just rise up and throw off the Jews—if we could just break their stranglehold on everything good—things would be okay.

This story is so familiar that some have grown callous to its absurdity. So, it may help to consider the conspiracy theorist David Icke, who has made his career arguing that we're all unwitting slaves of a race of alien lizard people hailing from the fourth dimension. We don't notice that all US presidents have been extraterrestrial lizards because they wear holographic veils that hide their onyx eyes and scaly tails. Icke claims to have found the fountainhead of all human suffering in the peculiar dietary preferences of our lizard overlords.

Lizards feed on human energy and find misery to be the sweetest chi. So, the lizards beset humanity with war, poverty, and disease in order to brim a trough with our sorrows that they can then lick up with their thin reptilian tongues. According to polls, about twelve million Americans believe this story is true.[5]

Now go back through the above paragraph and substitute the word "Jew" for "alien lizard person" and you pretty much have the story that mobilized not only the Tree of Life killer but the Nazis as well.* When the Nazis came to power in 1933, Hitler charged Joseph Goebbels with establishing the Reich Ministry for Public Enlightenment and Propaganda. The ministry brought Germany's entire apparatus of storytelling under state control in order to author a new national reality. On the radio, in the papers, in newsreels, in speeches, and even in manufactured whisper campaigns (*mundpropaganda*), Nazi propagandists appeared to tell many stories, but really they told only one—a story of Aryan knights battling Jewish evil in humanity's last, great stand. The story was so simple and powerful, so thrilling to the masses, that it made fiction real.[6]

The scale of suffering caused by this particular story surpasses comprehension. Anti-Semitism played a central role not just in the Holocaust but in the Nazi's overall justifications for waging World War II.[7] In addition to the murder of millions of Jews, the story contributed to tens of millions of additional war deaths, countless millions of mutilations and rapes, the obliteration of ancient cities, and the vaporization of immense material and cultural wealth. This is all tied to a story of evil Jews that's nearly as dumb as David Icke's lizard fantasy.

And now the ever-expanding death toll of this dumb story has expanded to include the one-man pogrom at the Tree of Life Syn-

* Some accuse Icke of laying out a sly allegory of the world Jewish conspiracy. In this accusation of anti-Semitism, Icke finds yet more evidence of the world lizard conspiracy's campaign to discredit him.

agogue. Here is the infamous final message the shooter posted to social media before opening fire: "HIAS [the Hebrew International Aid Society] likes to bring invaders in that kill our people. I can't sit by and watch my people get slaughtered. Screw your optics. I'm going in."

Here's what the killer was saying: I'm not a fool. I know it doesn't *look* good spraying bullets into a crowd of helpless, mostly geriatric worshippers. I know that I'll seem like a monster. But look closer. *I'm not the monster. I'm the good guy who sacrifices everything to slay the monster.*

This is revealing. It points to the deep meaning of what happened in that synagogue. The Tree of Life killer was no mere fanboy of ancient fictions about Jewish evil; at some point he entered the fiction as a character. He made himself the punishing hero of history's greatest epic. The killer had enmeshed himself in a nightmarish LARP (live-action role-playing) fantasy, like those grown adults who rush happily through the woods playing out *Dungeons and Dragons* fantasies.

But his victims were real.

The Essential Poison

In *The Storytelling Animal*, I argued that *Homo sapiens* (wise person) is a decent definition for our species. But *Homo fictus* (fiction person, story person) might be even better.[8] Human beings are storytelling animals.

But now I'd like to sharpen my description of *Homo fictus*, accentuating a unity in humanity's storytelling and tool-using instincts: Humans are the animal that uses story like a tool. A tool, broadly defined, is a device we use to influence and modify the world around us. As inveterate tool users, people use hammers to drive nails, screws to connect boards, syringes to save lives, computers to calculate and connect.

But we also come into the world endowed with natural tools. Our hands, for example, are all-purpose tools that we use to wave around and communicate, to show affection through caresses, to fabricate other tools, and to ward off aggressors.

Stories function as mental tools that we use not to modify the world around us so much as the people around us. A storyteller asks, How can I sway people? How can I get their money? How can I gobsmack them with the beauty of life? How can I convert them to my worldview? How can I make them laugh so they'll like me? How can I win their vote? How can I bring them together as a team? How can I change the world? And story—fiction, narrative nonfiction, and everything in between—is the natural lever used to move an audience into harmony with these goals.

Like all tools, stories have proximate and ultimate purposes. They do X *for* Y. A hammer is proximately for driving nails but ultimately for making something, say, a table. At a proximate level, stories also fill a variety of roles, including entertaining, teaching, and producing meaning. But these functions are part of the larger sway-making function of storytelling, not distinct from it. Stories are influence machines, with predictable elements designed to seize attention and generate emotion toward the ultimate end of gaining different types of influence over others.

It may help to think of the sway-making power of stories as the closest real-life equivalent to the force in *Star Wars.* Like the force, story is an all-pervasive field of dark and light energy that influences all of our actions. On the radio, on the news, on TV, on podcasts, on social media, in advertising, and in face-to-face yarning—we're forever swimming through a turbulent sea of narratives, with rival stories churning against each other and buffeting us around.

The force in *Star Wars* is, of course, a purely fictional concept, at the center of a fictional religion, at the center of a fictional universe. But the story force is real. It's a natural phenomenon that can be pinned down and studied scientifically just like any other natural force, from electromagnetism to earthquakes. In fact, over the last few decades an authentic science of storytelling has emerged to challenge the ancient superstition that there's something spooky about story—something that makes it science-proof. Today a broad consortium of researchers, including psychologists, communications specialists, neuroscientists, and literary "quants," are using the scientific method to study the "brain on story."

The results are frankly unsettling. They show that a master storyteller isn't unlike a master of the force. Storytellers cast spells that grant them entry into our minds, where they change what we feel, which changes how we think, which can change how we spend, vote, and care.

This is a wonderful thing when the story force's massive capacity for good is channeled by a true Jedi. But this sway-making capacity is just as available to the Dark Side.

Like the dualistic force in *Star Wars*, story science reveals that everything good about storytelling is the same as everything bad. Everything that makes storytelling wholesome is precisely what makes it dangerous.

So, yes, as I told the nuclear safety instructors, story is a precious tool for teaching and learning. But this also makes it a perfect tool for manipulation and indoctrination.

Yes, narratives are the primary tools we use to make sense of the world. But they're also our main tools for fabricating dangerous nonsense.

Yes, stories typically have moral dimensions that reinforce prosocial behavior. But in their monotonous obsession with plots of villainy and justice, they gratify and reinforce our instincts for savage retribution and moral sanctimony.

Yes, empathetic storytelling is among the best tools we have for overcoming prejudice. But it's also how we construct those prejudices, encode them, and pass them along.

Yes, there are countless examples of stories that have helped societies find their better angels. But throughout history it's always been stories that have conjured the worst demons too.

Yes, stories can act like magnets that draw hodgepodges of humanity into cohesive tribes. But stories are also the main thing pushing different tribes apart—like magnets turned the wrong way around.

For these reasons and more, I think of storytelling as humanity's "essential poison." An essential poison is a substance that's crucial for human survival, but also deadly. Consider oxygen. Humans, like all breathing creatures, need oxygen to survive. But oxygen is also a highly volatile compound (one scientist dryly calls it "a toxic environmental poison")[9] and, over our lifetimes, it does great cumulative damage to our bodies.

When oxidation hits your car, the vehicle turns rusty, falls apart, and dies. When oxidation afflicts the organic structures of the human body, you get snapped strands of DNA, stiffened arteries, fissures in cell membranes, and likely contributions to around two hundred degenerative diseases. The bodily damage caused by oxidation—oxidative stress—is also a primary contributor to the general process we call aging. When we begin to feel "rusty" in middle age, we aren't literally corroding, as metal does, but we're probably feeling the effects of oxidative stress. Researchers have called this "the oxygen paradox": oxygen is indispensable but also, in the end, quite harmful.[10]

This book is about the story paradox. Story is humanity's ancient curse and blessing. It's our disease and our cure. It's our doom and our salvation. Story raised us up as a species. Our narrative capacity helped a soft, weak, insignificant hominid gain dominion over the planet.[11] But now we're living through a big bang of storytelling—a shockingly rapid expansion of the universe of stories in every direction. We're living in the age of social media, peak TV, twenty-four-hour news channels, and skyrocketing total media consumption. The sudden evaporation of technological barriers to entry means that any person who wants a communications empire can compete for one. We can disseminate print, visual, and aural content *instantaneously* through a network with global reach—something that even major media companies couldn't match a couple of decades ago. And in this era of massive technological and cultural upheaval, story threatens to derange our minds, maroon us inside different realities, and tear our societies apart.

When I say stories are driving the whole species mad, here's what I mean. It isn't social media making us crazy and cruel, it's the stories social media spreads. It isn't politics severing us from one another, it's the wedge-shaped stories politicians tell. It isn't marketing driving us toward planet-killing overconsumption, it's the happily-ever-after fantasies that marketers spin. It isn't ignorance or meanness that leads us to demonize one other, it's instead a naturally paranoid and vindictive narrative psychology that leads us to become suckers over and over (and over) again for simplistic stories of the good guys fighting the bad.

My focus will be divided between spectacular examples of the dark power of stories and the quieter way that power influences ordinary people—every last one of us. The Tree of Life killer may have been born a monster. But more likely a story turned him into one. He illustrates one of the great laws of human history, and one of the constant themes of this book: Monsters behave like monsters all the time. But to get good people to behave monstrously, you must first tell them a story—a big lie, a dark conspiracy, an all-encompassing political or religious mythology. You have to tell them the kind of magical fiction that turns bad things—like wiping out world Jewry—good.

Behind all the factors driving civilization's greatest ills—political polarization, environmental destruction, runaway demagogues, warfare, and hatred—you'll *always* find the same master factor: a mind-disordering story. If *The Story Paradox* isn't a theory of everything in human behavior, it's at least a theory of the worst bits.

The most urgent question we can ask ourselves now isn't the hackneyed one: "How can we change the world *through* stories?" It's "How can we save the world *from* stories?"

1

"THE STORYTELLER RULES THE WORLD"

THE STORYTELLERS MAY RULE THE WORLD, BUT YOU WOULDN'T know that from visiting Washington & Jefferson College.* Even at this small liberal arts college, as at almost all other colleges and universities, it's obviously the scientists who rule.

This hit me when they tore down McIlvane Hall. A century's worth of Washington & Jefferson students and teachers, myself included, had gathered there to explore the mysteries of philosophy, sociology, and English literature. But one morning, a lone man rumbled up on the back of a big machine that had a long neck and powerful jaws. It looked like some kind of iron brontosaurus that groaned and farted as it stretched out its neck and tore down all that craftsmanship in huge messy bites, then spit it out as garbage on the ground.

They built the new Swanson Science Center in its place—fifty thousand square feet of brick and polished stone and laboratories gleaming with glass and steel. Inside, a huge marbled atrium soars

* "The storyteller rules the world" is a proverb of uncertain origin, variously attributed to authorities including Plato, Aristotle, a Hopi elder, and a Native American medicine woman.

sixty feet high, with grand pillars and Palladian windows stretching floor to ceiling. It's a Taj Mahal of science built at a cost of tens of millions of dollars.

I watched the new science center rise brick by brick from the back stoop of the English Department, not forty feet away. Davis Hall, where the English Department is housed, has a certain shabby grandeur. This nineteenth-century colonial-style home has high ceilings and hardwood floors and a stately porch. But get closer and you'll see that anything that can rust is rusting, anything that can peel is peeling. You'll see how rain sluices through the porch roof to warp the deck boards. You'll see the paint curling back from the softening shutters, showing green fuzz underneath. Look up and you'll notice vandalous saplings growing in the leafy mulch around the chimney, with their roots pushing through the crumbling bricks. Walk around back and you'll see surface rot eating its way into the building like melanoma.

When the science center was finished, I stood way back and took in the view of Davis and Swanson pressed close together like a great whale and its crusty little barnacle. I know a good metaphor when I see one.

Davis Hall slowly falling to ruin—the old religion of the humanities dying.

The Swanson Science Center—a temple to a new god risen.

I'm not a humanities zealot making a doomed last stand in the war between the "two cultures" of literature and science. That war is over. The culture of science won in a rout. The National Science Foundation budget for 2020 was $8.3 billion. The National Endowment for the Humanities got $162 million—more than fifty times less. And the movement toward STEM-focused education reflects a huge national bet, a unified societal consensus that the subject matter under the jurisdiction of university science centers is our future, and the subject matter under the jurisdiction of the world's Davis Halls is of the past.

But what if we're wrong?

What if stories have incredibly deep and powerful effects on our lives, but few of us actually know it? What if our storytelling psychology—the way our minds shape stories and are shaped by them—was not just a problem but the problem of problems? What if stories are the master factor driving so much of the world's chaos, violence, and misunderstanding? And what if there's something about the specific design of stories and the way they lock into our brains that makes this hard for us to notice?

In the Taj Mahal of science, they study the great prime movers of nature—the elemental physical and chemical rules of the universe. The Physics Department reveals the secrets to the atom, which holds the power to end civilization or to fuel its rise. Classes on neuroscience reveal the basic hardware and software of the human brain. Over in the college's temple of information technology, professors describe the evolution of a new race of AI angels and demons that may, one day, have the power to make heavens or hells for their creators.

In the offices of Davis Hall, you find the scholars who study how stories work, and you find the creative writers who teach young storytellers how to put this knowledge into practice. It's a building devoted to the study of the great prime mover of human nature—the story. If the storytellers actually do rule the world—if they actually do author our destinies—then the most powerful and volatile force on Earth is actually being studied in Davis Hall.*

The question is, Can we learn to control it?

Life in Storyland

My PhD is in literary studies, which means I'm trained less to appreciate the magic of stories than to figure out how the magic works

* After I submitted the second draft of this book to my editor, Washington & Jefferson College didn't install the English Department in a gleaming new Taj Mahal of the humanities. But it has thoroughly fixed up Davis Hall, inside and out. The old pile looks grand.

and pass the knowledge along to students. In recent years, I've taken a leave from the classroom to focus on writing. But back when I taught for a living, I liked to greet new students with a little quiz. The same goes, even today, when I'm invited to visit colleges and give guest lectures.

On a scale of 1 to 10, I ask the students, how important are stories in your life?

They scratch their heads. The question is too vague. So, I say a little more. In comparison to everything else in your life—your job, your religion, your sport, your family, your hobby, your boyfriend or girlfriend—where do stories rank on a ten-point scale?

A few hands come up. *Two? Three?*

But the students are hesitant. They want to work with me, but my question still seems too subjective. Okay, I say, let's make this concrete. The hours of your life are a precious and finite currency. How you spend that currency is arguably the most objective way of measuring how much you value one thing versus another. You spend a lot of time with your boyfriend because you value him so highly. You spend as little time as possible with your annoying neighbor because you don't.

Heads nod. The students are following me now. I go through a series of questions, asking for estimates of the time they spend in storyland: How many minutes per day do you read stories—everything from the books assigned in class to nonfictional news stories? How many minutes do you spend watching stories of any kind on TV—reality shows, sitcoms, documentaries? How many minutes do you spend bopping along to the rhythmic short stories of popular songs? How about story-based podcasts or audiobooks? How many minutes consuming the personal story feeds of celebrities and friends on Instagram or Snapchat? How many minutes do you spend inside narratives built by video game designers?

At the end of this highly unscientific survey, I ask the students to sum up their estimates and volunteer their totals. The last time I did this, some students seemed a little dismayed. They kept checking their math for errors. On average, they estimated that they spent more than five hours per day inside stories of various kinds. This was more time than they spent on any other activity. More time than they spent studying or practicing their religion or training for their sport or eating or hanging out with their friends.

By the economic logic of time spent, story didn't rank as a two or three in their lives. Story was a ten. It measured out as the most important thing in their lives.

When I run this little exercise, some students begin to slump in their seats, as if saddened by the waste of their lives. And they only slump lower when I point out that they've probably massively underestimated their actual story consumption. According to a recent Nielsen study, the average American spends almost twelve hours per day consuming media, including 4.5 hours of TV alone.[1]

Upon hearing this, the students' ego defenses kick in. Some question the Nielsen methodology. I point out that different researchers, using different methods, have reached similar results. Other students raise valid concerns about how I'm defining "storytelling" (does Instagram really count?) or question the whole notion that time spent is a good proxy for value. Once a student impishly volunteered that he'd been compelled to spend an outsized chunk of his life in English classes, and yet he didn't value the subject at all.

I tell the students that these are excellent points. But no matter how we define storytelling or measure our intake, we indisputably spend a huge portion of our lives consuming stories. Which brings me to the question I've been sneaking up on all along—a question so big and basic that most students have never given it a thought. Why? Why do people like stories so much? And weirdest of all, why do we care so much about fiction? Why do we hang on the fake struggles of pretend people? Wouldn't most of that time be better invested in studying? Or spending quality time with our loved ones?

Or doing some charity work? Or trying to attract a mate? Or finally learning that second language or musical instrument?

The students stare down at their desks, turning the question in their minds. The silence stretches out until someone raises a hand.

"Escapism?"

Everyone nods.

Why do people like stories so much? They're fun. They're candy.

Escapism is our culture's main theory of the purpose of stories. Because stories can feel like a pleasurable escape, we think that must be what they're for. But we shouldn't let the pleasure fool us. Stories aren't inert in their effects. We consume stories gluttonously throughout our lives, and we become what we eat. Let me illustrate with a story of my own.

The Girl in the Woods

The girl stood in the closet breathing so hard. So loud.

Oh god oh god oh god. Stop my breath stop my breath stop my—

The big man was in the room now, scuffing past the closet in his boots. He dropped heavily to one knee and peered under the bed. Standing, he yanked up his sagging jeans. He turned and stared at the closet door. He took a step closer. And then another.

The girl held her breath. But her heart was thudding so hard. So loud.

Oh god oh god oh god. Still my heart still my heart still my—

The man was standing very close.

The girl was standing very still.

The man leaned in and sniffed at the slatted door, raising his chin a little with each sniff. And then he smiled and showed his teeth. "Oh darling," he sighed.

She threw her shoulder against the door, drove into it with everything she had.

The big man sat back heavily on the bed, hand to his nose. His eyes were held wide and dark blood was pouring through his fingers.

For a moment she stood frozen, panting.

And then she was flying.

She heard him curse. Heard his boots squeak and scuff as he rose to give chase.

She flew down the stairs without touching them with her feet. She leaned and swooped through the open front door like a bird from a cage.

The smell of autumn leaves. The smell of damp earth. The moon visible through naked limbs. Soon she couldn't hear him crashing through the brush wailing her name.

The sound of her breath. The sound of the wind.

The sound of dry leaves shattering beneath her feet made her sad. When she heard the sound, she knew she wasn't flying anymore.

But she wasn't sad for long. She didn't need that magic anymore. She just needed to run like she knew she could. To run until she found it. Whatever it would be. A road full of cars. A river she'd follow to a little town. Or a single house with good people inside.

She ran fast and still faster, leaping logs, splashing through mud, her long hair streaming like a pennant.

She began to tire. She stumbled and fell. Her thin legs were all blotchy from the cold, shaking with fatigue. She felt the pain in her bare feet, saw that they were filthy with blood and dirt.

She was afraid again. Her strength was almost gone. And where was the road full of cars? The house with good people inside?

A shadow loomed out of the trees.

Let him be the good guy, she prayed, *the one I've been running toward.*

But then she saw the size of him. Saw the front of his T-shirt V'd with blood. He was creeping forward slowly, hunching as if he hoped to seem small.

"It's okay," he was saying. "It's okay."

"No, no, no, no, no," she said. She tried to take flight, as she had before. But her heart was too heavy. So, she only limped away, sobbing. She could hear his voice getting louder, his footsteps getting closer.

"No, no, no," she said, turning to face him. He inched forward with his palms out. She held his eyes while sweeping the ground with her foot, searching for a rock that would fit her hand just right . . .

You're the Girl

A few days ago, the girl in this story came to me for no reason at all. I just saw her in my mind, sprinting through a dark forest, and I began to wonder why she was out there all alone. My mind flashed to a different scene. I saw the big man scuffing across the bedroom. Saw him *through her eyes*, as though I were the girl hiding in the closet.

I paced around my basement office dictating images into my phone. And then, yesterday, I roughed out the scene in draft. Today I spent an hour liposuctioning my fatty first-draft prose, busting long sentences to shards, trying to create a breathless feel. It's mostly finished now. But before anyone else sees it, I'll go over it at least twenty or thirty times, building it up word by word, then shaving it back down.

If the girl in my story were real, she would be experiencing an intense fight-or-flight response. In the closet she'd be breathing hard to charge her blood with oxygen, and her heart would be racing to speed that blood to her muscles. Her brain would be pumping out a cocktail of hormones that would, among other things, facilitate faster blood clotting and increased pain tolerance. As the big man approached the closet, her pupils would dilate and she'd stop blinking. At the peak of danger, as the big man sniffed at the door, her vision would tunnel and she'd lose her peripheral vision.[2]

This powerful physiology of fight or flight explains how a small girl could plausibly knock back such a big man. It explains how she's able to run so fast and why she doesn't at first feel the frigid air or the rocks slicing into her feet. It explains why she feels so high when she first escapes—high enough to imagine the sensation of flight.

Now imagine that the girl in my story is fleeing for her life across the pages of an unputdownable novel. Imagine that you're on your couch reading the novel, with your fingernails between your teeth. If a team of scientists snuck into the room and hooked you up to the right machines, they'd find that your heart rate and blood pressure were elevated, just like the fictional girl's. You'd be breathing faster, too, and the galvanometers would measure an uptick in your perspiration. Just like the girl's on the page, your brain would be soaking your synapses with adrenaline and cortisol. Amazingly, your endorphin system would be revving up, pumping out endogenous opioids that would measurably increase *your* tolerance for pain. Irrespective of the light conditions in your living room, when the fictional danger peaked, your pupils would dilate, you'd stop blinking, and when the scientists waved their hands on your periphery, you might not notice. And all the while, your brain would be lighting up, not like someone who was passively *observing* a girl in danger—it would light up as though *you* were the girl in danger.[3]

The Storyteller

In 1947, Nat Farbman spent weeks hiking the Kalahari Desert photographing Khoisan hunter-gatherers. The photo essay he published in *Life* was a celebration of human diversity but also of shared humanity.[4] The photos capture the Khoisan's simple satisfaction in

CREDIT: Getty Images/ Nat Farbman

work, the tenderness of their family bonds, the gleeful romping of their children, and above all else their love of story.

Farbman snapped three photos of a Khoisan elder spinning a tale. The most famous is simply called *The Storyteller.*

Humans live in a storm of stories. We live in stories all day long. We dream in stories all night long. We communicate through stories and learn from them. Without personal life stories to organize our experience, our lives would lack plot and point. We are storytelling animals.

But why?[5]

Evolution shaped the mind *for* story, so it could be shaped *by* story. Story originated as a means for preserving and passing information about everything from religious and moral imperatives to specific advice about hunting or marriage. Cultures are big, complicated mechanisms. "The problem," writes the neuroscientist Antonio Damasio, "of how to make all this wisdom understandable, transmissible, persuasive, enforceable—in a word, of how to make it stick—was faced and a solution found. Storytelling was the solution."[6]

Take our Khoisan storyteller. He's gathered the band's children to tell them a tale about Jackal, the great trickster. You can see from their faces that the tale is great fun, but it's also, as the *Life* article puts it, a source of instruction: "When night comes and they [the children] sit around a fire with their elders, listening to an old man's stories, they begin to understand that the band is a unified group to which they belong and without which they cannot live." The Khoisan elder might have given his lessons straight—the hunter-gatherer equivalent of a bullet-riddled PowerPoint. But he knew in his bones what science has only recently confirmed. We learn most and best through stories. In a deep sense, this is what stories are *for*: grabbing us, teaching us, and influencing how we interact with the world.

The benefits of story go both ways. Storytellers give us a lot, and we give back amply in return. Anthropologists find that tribal storytellers all around the world enjoy high social status. For example, a recent study in *Nature Communications* found that, among Agta hunter-gatherers in the Philippines, good storytellers are lavished with benefits.[7] On average, they receive more resources, achieve better mating success (measured in number of children), and gain higher popularity within the group. Even though the Agta depend on skilled fishers and hunters to survive, they revere storytelling ability over all other skills. In short, the Agta prefer a reliable giver of absorbing stories to a reliable getter of life-giving meat.

It's the same for us. We have an insatiable appetite for quality stories and we lavish rewards on the people who tell them. Some of the most revered, high-status members of our society are our fiction makers—our star writers, filmmakers, actors, comedians, and singers. The *Forbes* list of the world's most highly compensated celebrities is dominated by these types of story artists, with athletes placing second.[8]

This is fairly weird. Like the Agta, we don't bestow fame and fortune on the people who keep us alive—the doctors who cure our illnesses, the sanitation engineers who keep us from getting sick in the first place, the government officials who make our society run, the farmers who feed us, or the soldiers who protect us. But we do lavish inordinate fame and fortune on the masters of make-believe—people who spend their lives doing the equivalent of playing publicly with dolls.

The Media Equation

When my eldest daughter was about three years old, she developed a straightforward theory for the miracle of television: it looked like tiny people lived inside the TV because *tiny people actually lived inside the TV*. Lots of preschoolers form the same theory. If you show them an image of a heaping bucket of popcorn on the TV screen, then ask what will happen if the whole TV is flipped over, most say the popcorn will spill.[9]

"Oh, how cute!" we think to ourselves. "The adorable little monkeys haven't yet learned that TV depicts representations, not realities." But when it comes to confusing representation with reality, we're all deeply confused monkeys.

Grown-ups know that there are no people living inside the TV. But we still get scared watching a horror film as if the serial killer were real. We get misty at a sad film as if there were a real fiancée and she really died. These brain processes are so ancient and deeply ingrained that we can't be educated out of them. My daughter eventually learned that the TV wasn't full of tiny people. But that hasn't stopped her from having nightmares after watching a scary film now that she's a teenager. Two media researchers from Stanford University, Clifford Nass and Byron Reeves, call this deep confusion of media with reality "the media equation," and it's easy to grasp even if you stink at math. Here it is:

Media = Real Life

Nass and Reeves note that the human brain didn't evolve to cope with an environment saturated with realistic simulations of people and things. Our brains completed most of their evolution back in the Stone Age, when there was no photography, film, or Dolby Surround Sound. So when we see convincing images of humans or convincing simulations of human life in stories, our brains reflexively process them just like the real thing. But there's more to it than that because, according to Nass and Reeves's data, people are nearly as bamboozled by purely text-based and oral forms of storytelling. Humans have been storytelling animals at least since behaviorally modern humans emerged around fifty thousand years ago. Our tendency to get swept up in stories as if they were real cannot, therefore, be wholly ascribed to a mismatch between Stone Age minds and modern entertainment technology.

Nass and Reeves published *The Media Equation* in 1996, and in the quarter century since, the data in support have grown much stronger. Take the well-studied phenomenon of parasocial interaction—the ubiquitous tendency to react to story characters in the same natural way we react to actual human beings.[10] The psychopathic crimes of Joffrey Baratheon in HBO's *Game of Thrones* fill us with disgust, and we vent our spleen online, as though we didn't know that Joffrey is just a young man named Jack Gleeson who's good at *acting* cruel.

And, unfortunately for Gleeson, it doesn't end there. In studies, people assume that actors' personalities resemble their characters', even when psychologists take pains to assure them they don't.[11] Of course, at a conscious level, everyone knows that actors are just playing elaborate games of make-believe. But the deep, dark parts of our brains can't unlearn what the story has taught them. (Thus, typecasting.)

I don't want to exaggerate. When you read or watch a scary story, your heart doesn't pound as hard as it would if you were actually in fear for your life. But it would be equally misleading to undersell story's emotional and physiological power. Fiction is fake. The monsters are fake and the wounds are fake. But they leave real scars. When scientists asked people to describe something in the media that traumatized them, 91 percent of people volunteered fictional horror stories rather than footage of real horrors such as mass shootings or the 9/11 attacks. The horror stories provoked symptoms reminiscent of PTSD, including intrusive thoughts, insomnia, and fear of being alone. For many people, the anxiety inspired by films like *Jaws* and *A Nightmare on Elm Street* persisted for years, with 25 percent reporting lingering effects six years later.[12]

Nass and Reeves's media equation is connected to narrative transportation—possibly the most important concept in story science.[13] Narrative transportation is that delicious sensation of opening a book, or turning on the television, and mentally teleporting out of our own mundane realities and into alternative story worlds. When transported, we partially decouple not only from the real world *but also from ourselves.* We identify so strongly with the protagonist of a good story that we leave our personal baggage behind—our preconceived notions, our dumb prejudices. And we're able to see life from the perspective even of people very unlike ourselves. This is what makes story such a potent driver of change. Minds that are closed in real life swing open in storyland.

Which brings me back to the girl who's still standing out there in the woods, eyeing the big man across the clearing. Now that I think about it, I don't know for sure how the girl found herself in that closet. And I have no idea what will happen to her next—whether she's a kickass heroine or a distressed damsel, whether she's in a thriller or a tragedy. I know only one thing for certain: I dreamed the girl to life and I'm watching over her. I'm the god of her little world. And because I'm all-powerful in my realm—and mostly benevolent—the girl will encounter no monster she cannot face down, no trauma she cannot surmount.

But no matter how I decide to shape the girl's story, if I'm able to transport you to those woods, I hold you in my power. The stronger the transportation—the more vividly you can feel the fear, and see the blood, and hear the crunching leaves—the stronger my grip. If this story were about a young woman menaced by an escaped mental patient, I could make you more supportive of harsh sentencing of criminals with mental illnesses. I know this because, in a pioneering study of narrative transportation, the psychologists Melanie Green and Timothy Brock used just such a story to produce just such an effect.[14]

On the other hand, maybe my story of a girl in peril isn't as cliché as it seems. Maybe I'm just setting you up for a plot twist: it's the girl who's mentally ill (she seems to think she can fly, after all), not the man. In the depths of a dangerous psychotic break, the girl can't see that the big man is her own loving father, who's trying desperately to save her. If I told *that* story well, I could dial up your support for government programs friendly to people with mental illnesses and their struggling families.

But if I wanted to influence mental health policy, wouldn't it be a lot simpler and faster to jettison the story and just come out with the facts and the arguments? Absolutely. But that's often less effective. We consume fact-based arguments with our defenses on high alert. We're critical and suspicious—especially when those arguments run counter to our existing convictions. But when we're absorbed in a story, we relax our intellectual defenses. As the narratology researcher Tom van Laer and his colleagues put it, after analyzing every relevant study in the science of stories, "narrative transportation is a mental state that produces enduring persuasive effects without careful evaluation and arguments."[15] In other words, strong storytellers do end runs around the brain's processes for sifting and evaluating claims. They can implant information and beliefs—often quite strong ones—without any rational vetting.

But what's a story, anyway? At a good library you'll find shelves sagging under the weight of maddeningly arcane attempts to draw a wavering line between what counts as a story and what doesn't. For the purposes of this book, I'm bypassing all of that in favor of a sequence of metaphors that convey what stories are *like* (the force, the essential poison, and more to come) paired with a basic commonsensical definition. In broad terms, a story or narrative—I use the terms interchangeably—is simply *an account of what happened*, whether it's a description of what happened in the real world or in a toddler's make-believe sessions.

Here's the sort of story this book *isn't* about: "I woke up this morning, went to the store to buy bread, and then ate the bread while reading the newspaper." This is what might be called *transparent narrative*—an attempt to efficiently transfer information out of one head and into another. Transparent narrative has no special power to sway.

This book is about what might be called *shaped narrative*. Shaped narrative conforms to a strongly stereotyped structure that I'll describe in detail in Chapter 4 and applies regardless of where the story sits on the fact–fiction continuum. For now, it's enough to say that shaped narratives typically focus on the struggles of protagonists, are almost always grounded in some implicit or explicit moral conflict, and ultimately lead not just to an account of what happened but also to *an expression of what it all meant*. Shaped stories are meaning-making tools and they can exert awesome sway, not just over individual people but also over whole civilizations.

Fake Gay, Black, and Muslim Friends

In a modern world grown clamorous with billions of people vying for attention in millions of stories, the proverb "the storyteller rules the world" needs a tweak. Everyone is telling stories. It's the *best* and most transporting stories that rule. And for around a century, the world's most talented storytellers have congregated in one place, Hollywood, California. Gradually, they've remade the world.

Or so argues Ben Shapiro in his 2011 book *Primetime Propaganda: The True Hollywood Story of How the Left Took Over Our TV*. The book has an arresting premise: television is "a leftist oligarchy" where an overwhelming preponderance of the writers, producers, studio honchos, directors, and actors are liberals. Together, they have gradually, but radically, changed American culture, pushing us away from traditional values of American exceptionalism and Judeo-Christian morality.

When I opened Shapiro's book, I expected a diatribe. Shapiro is an adamant conservative partisan and his main point sounds a lot like the paranoiac fantasies Alex Jones shouts about on InfoWars: "The box in your living room has been invading your mind, subtly shaping your opinions, pushing you to certain sociopolitical conclusions for years."[16] In other words, Shapiro suggests, a tiny cabal of elite artists and their limousine-liberal patrons are attacking the rest of us with an unprecedentedly "impressive weapon" of thought control.[17] But Shapiro doesn't allege any kind of conscious lefty plot. His point is that modern Hollywood is ideologically all but monolithic, and its stories can't help but reflect this fact.

We think of America as being eternally split down a Left–Right continental divide. But though this may be hard to believe given the recent tenure of the forty-fifth president, America has liberalized massively in recent decades. Self-described conservatives today generally hold more liberal views than self-described liberals did little more than a generation ago.[18]

This steady leftward march—and the reactionary spasms it has so far trod down—owes to many sources, but I'm inclined to agree that Hollywood storytellers deserve much of the credit. Or blame. Depending on your personal politics, lefty predominance in Hollywood has either cohered America around wholesome ideals of diversity and equality or it has set us awash in culturally corrosive, if pleasurable, brainwashing.

It's worth quickly noting that left-wing intellectuals are also aware of the dangerous power of Hollywood. Only, whereas conservatives like Shapiro worry about losing a culture war in the West, the political Left worries about *winning* on a worldwide scale. From the Left's point of view, the arsenal of the American empire encompasses the soft power of our popular storytelling just as much as our bullets and bombs. Even in the most remote outreaches of planet Earth, most people spend big chunks of their days inside a virtual America—inhabiting the stories we pump into their radios, TVs, and theaters. Leftist thinkers see this as a bloodless imperial conquest that has gently subjugated indigenous cultures and forced the entire world to think, speak, dress, and buy like us—to adopt our unrealistic beauty standards and our soulless materialism. America, they suggest, was the first nation to achieve the goals of empire largely through a conquest of popular art.

Shapiro's book is based on a solid plausibility argument. He didn't consult the science on the persuasive power of storytelling. If he had, he could have made an even stronger case. Consider the rapid evolution of American views on same-sex marriage. Since 1996, popular support for same-sex marriage has soared by forty percentage points.[19] The swiftness of this change flummoxed social scientists, who expect much slower erosion of entrenched cultural biases. On a 2012 episode of *Meet the Press*, Vice President Joe Biden surprised viewers not only by endorsing gay marriage but also by attributing historic changes in American views on homosexuality to the sitcom *Will and Grace*.

The Daily Show and *The Colbert Report* had a field day, with John Stewart breaking into a show tune about "fictitious gays" and with Colbert working the statement into a montage of Biden gaffes—alongside the vice president's claims that "jobs" is a three-letter word and that only people with Indian accents can shop at 7-Eleven. Admit-

probably does help but only to bring marginalized people up to a pre-existing bourgeois standard. they become "normal" tedly, the claim that a mere sitcom helped drive a massive and ongoing social transformation seems preposterous at first blush. But Biden was actually expressing the gist of a leading social scientific theory.

Here's how the theory goes. Research shows that regular contact with gay friends or family members is a better predictor of gay-friendly attitudes than gender, level of education, age, or even political or religious affiliation. And the same seems to be true for the illusory relationships we form with fiction characters. We relate to the characters on the sitcom *Friends*, in other words, as though they were our real-life friends. And the relationships we form with these characters can be so authentic seeming that it hurts us when they end. When *Friends* ended, many fans of the show suffered the kind of distress associated with breakups of real-life friendships, with the strongest effects for the loneliest viewers.[20] This tendency to form authentic-seeming connections with media personalities is surprising until you recall how we spend our leisure time. We spend hours per day absorbed in virtual social interaction with TV characters, while only averaging around about forty minutes socializing with family and friends.[21]

When we're absorbed in fiction, we form judgments about the characters exactly as we do real people, and we extend those judgments to generalizations about groups. When straight viewers watch likable gay characters on shows like *Will and Grace*, *Modern Family*, and *Schitt's Creek*, they come to root for them, to empathize with them—and this shapes their attitudes toward gay people in the real world. Studies indicate that watching television with gay-friendly themes lessens viewer prejudice.[22]

The implications of these studies extend beyond the issue of homosexuality and bias. For example, if you're white and you make a black friend, studies show that any prejudice you harbor toward black people will likely diminish.[23] And the same thing happens if you make faux black friends with the likable characters of *Black-ish* or *Black Panther*. Similar effects have been shown for TV shows featuring Muslim leads or characters with mental illness. Most encouragingly, these effects seem to be more robust and long-lasting than standard approaches to prejudice reduction, like diversity training, which hasn't been shown to work as well.[24]

All of this research adds up to a surprising possibility: fictional characters from Will and Jack, to the big-screen characters played by Sidney Poitier and Viola Davis, to the protagonists of novels like *Roots* and *Beloved* may have improved the lot of American minorities as much as direct political action has.

Put another way, new research doesn't just suggest that fictitious gays blazed the trail that led to Barack Obama's historic endorsement of gay marriage in 2012. It suggests that but for "fictitious blacks"—from Kunta Kinte to the powerful African American commander in chief on the television show *24*—we might not have had a President Obama in the first place.

The Great Loquacity

According to the English poet and philosopher Samuel Taylor Coleridge (1772–1834), enjoyment of fiction requires a "willing suspension of disbelief"—a conscious decision.[25] We say to ourselves, "Well, I know this story about Beowulf battling Grendel is hooey, but I'm going to switch off my skepticism for a while so I can enjoy the ride."

But that's not how it works. We don't *will* our suspension of disbelief. If the story is strong, if the teller has style and craft, our suspension of disbelief *just happens to us.* Think of the metaphors we commonly use to describe what stories feel like. Narrative transportation is always something done *to* us, not *by* us. It's a force we're subject to, not something we control.

We think of storytellers as metaphorical bruisers who overpower us and hold us down—they *hook, grip, rivet,* and *transfix.*

Or we think of them as metaphorical Svengalis—they *hypnotize, engross, mesmerize,* and *entrance.*

Or storytellers are lovers—they *infatuate, besot, intoxicate,* and *seduce.*

Or storytellers are forces of nature—they're rivers or winds that *sweep us up* and *carry us away.*

Or storytellers are powerful witches—they *enchant, charm, bewitch,* and *spellbind.*

We reach for metaphors when we lack language for direct description. We say not that story *is* X but that it's *like* X. But in the above collection of metaphors, we get a good sense for what story actually *is*: it's a drug. To see what I mean, consider one of my favorite studies encapsulating the tragicomic lovability of the human animal.[26] Researchers from Harvard and the University of Virginia brought subjects into a lab where they had to choose between two torture devices. The first option was to push a button that would deliver a safe, but still sharply unpleasant, electric shock to themselves. Two-thirds of the men in the study chose to shock themselves despite the fact that they'd (1) all been shocked in an earlier phase of the study, and (2) all professed that they would pay money to avoid the unpleasant experience in the future.

So why did these men, all of whom disliked the zap, nonetheless zap themselves again? It was because the alternative torture device, a simple chair, seemed even worse. Research subjects would have to sit in the chair for fully ten minutes and simply think. They were to do nothing else. The room was empty. They couldn't check their email, they couldn't swipe at their phones, they couldn't talk to anyone, and they couldn't even read the back of a cereal box. They just had to sit there thinking whatever they wanted to think.

The horror, the horror, the men must have thought, as they lunged for the button that would shock all thought from their brains.

Women took part in the study, too. But just (*just?*) 25 percent of them preferred painful zaps to sitting still and thinking.*

A few pages back, I waved off the escapist theory of storytelling a little brusquely. But now I'll acknowledge that escapism, though not an adequate theory of the attractions of storytelling, is a big part of it.[27] Psychological research demonstrates that our minds are constantly wandering—with a relentless voiceover narration. I call this inner voice "the Great Loquacity." We think of the Great Loquacity as synonymous with ourselves, with our own egos, even though we have little control over what it says.

Further, as the experiment with DIY electroshock therapy suggests, most of us aren't so fond of our own endlessly talkative company. The Great Loquacity may be our constant companion, but it's also our tormenter. Studies show that when the voice is maundering along in our heads, we aren't as happy as in those blessed moments when we can finally shut it up—even if it means short-circuiting the voice with little stabs of electricity.[28]

Mind wandering, of course, is nothing like all bad.[29] But when we chase our bliss, we're mainly chasing after experiences that will shut down the Great Loquacity. States of pleasure are, for us, virtually synonymous with the temporary muzzling of the inexhaustibly carping inner monologue. The things we pursue to decrease psychic pain and enhance pleasure—sex, movies, absorbing conversation, sports, recreational drugs, video games, mindfulness meditation, TikTok trances, and flow states of any kind (the feeling of being "in the zone")—are pursued in large part because they provide a temporary furlough from the prison cell of our own skulls, that feeling of being locked up for life with a celly who *Just. Won't. Shut. Up.*

So, stories do provide a kind of escape. But the escape they provide isn't only from our personal problems or the problems of the

---

* The researchers speculate that the gender split may reflect the fact that men, contrary to popular belief, have higher average thresholds not only for risk but also for pain.

world. It's more profound than that. They provide an escape from ourselves. Narrative transportation is valued not just because of where it takes us *to* but also because of where it takes us *from*: our own tedious company. The gift storytellers give us, as Virginia Woolf observed, is "the complete elimination of the ego."[30]

When absorbed in a good story—when we watch a favorite TV show or read a thriller we can't put down—our hyperactive minds go quiet and pay close attention, often for hours on end. It takes meditators years of dedicated training to even approach the level of mindful presence that we all naturally achieve by simply listening to *Radiolab*.

Of course, story doesn't count as a drug if you narrowly define drugs as chemical substances. But stories are still something we "take" to literally change our brain chemistry.[31] Like other drugs, we use story to relieve our pain, loneliness, and drudgery. And just like other drugs, story lulls us into an altered state of consciousness that bears comparison to a hypnotic trance—or even a psilocybin trip. When transported into the hallucinatory realm of story, our minds empty, our endlessly burbling streams of consciousness go still, and time races by. When the "active ingredients" of a story[32]—the plot, theme, character, style, and so on—are expertly compounded, we slide into highly receptive brain states that bring us under the storyteller's power.

This is lovely. All of the good things story does flow from the psychoactive properties of the drug—from the sway it generates.

This is scary. All the bad things story does also flow from the sway it generates.

2

THE DARK ARTS OF STORYTELLING

Plato stood on the Acropolis at dawn, with the Parthenon looming at his back. Far below he saw the city walls sprawled out in their rubble. He saw the fields and harbors where the battles had raged, the countryside the Spartans had ravaged. He could still make out the charcoal scars on the earth, where, when he was a boy, they'd burned plague corpses by the thousands.

He'd been born into an epoch of carnage. Born into a sickness that swept through Athens like an army, killing every third person. Born into a war with Sparta that lasted thirty years—*thirty years*—and knew no decency, no restraint.

Plato absently traced the scar that cut through his beard down his neck. He'd faced the terror of the Spartan phalanxes and lived. But would he survive this? The crazed violence of his own people?

Athens had been vanquished, lost its empire, and the Spartans had entered the city, red cloaks streaming. Every Athenian man feared the sword—every woman feared slavery and rape. But the

Spartans only installed a government of rich traitors who favored Sparta and all her ways.

And then came the time for brother killing brother, and sons slaughtering fathers. The Spartan puppets became known as the Thirty Tyrants. They terrorized the city. They killed and banished thousands until the people rose up and drove the Spartans out. But the people killed so many men in reprisal—not just tyrants but foolish men who spoke loosely or wore their hair in Spartan fashion.

The people even turned against Plato's teacher, Socrates. At the trial they called him a menace to society—a corrupter of the youth. Some cried out that the charges were baseless. But Plato knew better. The old philosopher—with his shoeless feet and false modesty, with his dingy tunics and never-combed hair, with his thin arms and his belly made fat with mooched wine—was the most dangerous man alive.

Plato checked the sun, which sat one finger above the sea. The old man would be down in his cell now, brewing his hemlock tea. Hemlock was better than the crucifix, but it was no gentle death. Nor quick. The old man would spend his last hours moaning and vomiting, then thrashing and gasping as his windpipe swelled shut.

Plato stared out over the sea, cursing the playwright Aristophanes. His comedy about Socrates, *The Clouds*, turned the people's hearts—taught everyone that the old man was a master of devious sophistry, not true philosophy. He smiled and shook his head, remembering how dearly the old man had loved that play. Whenever it was staged at some festival, Socrates would elbow his way to the front of the crowd, saying, "Pardon, pardon." He'd stare in delight as an actor costumed in filthy rags and a ratty wig raced around stage waving his arms and talking nonsense. Socrates would laugh until he hunched over in pain.

Plato set his broad shoulders and started down the steep path leading out of the city, ready to knock down any man who blocked his way. Even now a mob might be forming up at his home, hoping to drag him from his bed without a fight. He would hurry to Megara, where many of Socrates's followers had already fled to wait out the danger.

As he walked through the countryside, he turned a question in his mind like a potter spinning clay on a wheel. He could see the future. The Greeks would keep warring against themselves—blithering madly as they punched their own noses and stabbed their own guts. It would go on until they devoured all their strength, exhausted all their blessings. And when they were weak enough, some army would march down the peninsula to turn them all to corpses or slaves.

Wasn't there a way to live in peace and sanity, not murder and madness? A way to first wipe clean the slate written with the bad old ways, then chalk out a plan for a *Kallipolis*—a beautiful city—built on logic?

Maybe, he thought, the first step would be to find Aristophanes and hand *him* a cup of hemlock tea. Plato shook his head ruefully, dismissing the thought. The old man had loved storytellers so much—and feared them, too. He'd once told Plato that every person was a sad mishmash of all the stories they'd been told, all the lies passed down by grandmothers and priests, poets and tyrants. "No matter who sits on the throne in a city," the old man had said, "storytellers truly rule the world."

Plato walked faster. To make a perfect city, he realized, it wouldn't be enough to declare war on the likes of Aristophanes. To make a perfect city, you'd have to make war on all the storytellers—on the whole *phenomenon* of story.

Around the age of fifty, having long since returned to Athens and established, in his Academy, the first prototype of the modern university, Plato finished his greatest and most influential book, *The Republic.*[1] It was a how-to guide for creating a utopia founded not on the superstitions passed down by storytellers but on the purified logic of philosopher kings. It's widely considered to be the most ruthless insult against storytelling ever conceived by a great thinker.

Step one toward utopia: banish the poets, every last one of them.*

"Nothing Is Less Innocent Than a Story"

When modern people first hear that Plato so feared poets that he wanted to banish them from society, their first response is usually: "*What?*" Because, for most of us, what could be weaker and more inconsequential than a poem? So here's the first thing to get straight: when Plato speaks of "poets," he means fiction makers of any kind. In Plato's day, most stories, whether enacted onstage or read from the page, were told in verse. And then, as now, storytellers wove pointed social and political messages into their tales.

As an ultra-rationalist who yearned for a utopia ruled by philosopher kings, Plato condemned storytellers as professional liars who got the body politic drunk on emotion. Crowds slurped up the juicy stories—all the sex and violence, all the laughs and tears—while also slurping up immoral and dangerous ideas. Whatever the merits of his case, Plato, in his sometimes Aspergerish way, underestimated storytelling's grip on our species. Storytelling instincts are embedded deeply in our brains. The only way to "banish" them involves a skull saw and a scalpel.

But we shouldn't let the incandescent wrongness of Plato's solution distract us from the real problem he describes. We have a huge blind spot when it comes to stories. We think they're weak when

---

* Little is known about Plato's life and personality, though Greek tradition did portray him as a burly ex-wrestler and a combat veteran. The exact date of the appearance of *The Republic* is also unknown, though it is traditionally given as 380 or 381 BC. I invented the line "No matter who sits on the throne in a city, storytellers truly rule the world." However, it runs parallel to views expressed in *The Republic*, which may be why the proverb "the storyteller rules the world" is often falsely attributed to Plato or Socrates. The blank slate analogy is employed by Plato in Book VI of *The Republic*.

they're strong. We think they're frivolous when they're serious as hell. We think they're innocent when, as the narratology professor Tom van Laer and his colleagues put it, "nothing is less innocent than a story."[2]

Plato isn't a philistine. He isn't dead to the joys of story, nor does he deny story's many beneficial effects. He just asks a question that's as radical today as it was twenty-four centuries ago: Could it be that even the grandest specimens of storytelling—from the Homeric epics, to great stage plays, to fundamental religious myths—do more harm than good?

History has loved and hated Plato's *Republic*, but it's never quite been able to take his point seriously: beneath humanity's hardest problems are the illusions foisted upon us by storytellers. The closest Plato gets to approval is from modern scholars, who assure us that such a smart guy couldn't really have believed something so dumb.

Here's my view: Plato's solutions to the problem of storytelling (banishment was just one of the possibilities he tried out) range from bad to worse. But he was even more right about the problem than he knew. For one thing, Plato was writing only about fiction—about the volatile power of poetry, drama, and myth. But it's now obvious that the dominion of story is much, much vaster than tales of invented characters embroiled in pretend adventures. A story is simply an especially engaging way of structuring information, no matter whether the information is factual (as in a documentary or historical narrative), purely imaginary (as in a video game plot), or somewhere in between (as in a "grand narrative" like Marxism).

And now, the dominion of story in human life is growing all the time. We are a species with a bottomless appetite for story, and technology has broken down all barriers to how much we can make and consume. Alongside the technologically driven big bang of storytelling has come a parallel expansion in the scientific understanding of how story works on our minds. The main powers in our world have absorbed this science and are putting it to use.

Big business, for example, is all-in on storytelling as a tool of persuasion, and the world's great and rising powers are deploying stories in ever more devious and effective ways.[3] As technology has made conventional war so costly in treasure and blood, the battle-space is shifting away from the real world to the terrain of the human imagination.[4] Chinese military planners understand that dominating the battlespace of storyland is a crucial continuation of the goals of warfare—just by other means.[5] The ongoing Russian influence campaign will go down in history as the first use of doomsday weapons of story warfare. And the Pentagon is funding basic research on stories, with the goal of creating narrative Death Stars that could make the old propaganda methods seem as primitive as flintlock muskets.

Story was always strong. Story was always dangerous. But for all the good stories do, and all the comfort they give, our technology has made them more ubiquitous, more powerful, and more weaponizable than Plato could have dreamed.

Hidden Persuasion

In his 1957 book *The Hidden Persuaders*, the social critic Vance Packard alleged that an elite cadre of ad men and political operatives had teamed up with equally elite teams of social scientists and psychotherapists. Together, they were discovering the hidden keys of unconscious motivation. Those who possessed the keys could skip the arduous processes of persuasion and plant subliminal yearnings deep in the unconscious.

Packard's sensational allegation of a massive mind-control conspiracy would go on to sell millions of copies. It did so partly because of a bespectacled and bookish-looking market researcher named James Vicary.[6] Soon after *The Hidden Persuaders* hit bookshelves, the forty-two-year-old Vicary found a way to rescue his struggling marketing consultancy. He famously lugged a specialized projector called a tachistoscope into a New Jersey movie theater. While theatergoers enjoyed the William Holden film *Picnic*, Vicary strobed secret messages on the screen, but too rapidly for conscious minds to detect.

One message read "Eat popcorn." The other, "Drink Coke." Compared to a control group, Vicary reported, the 45,699 viewers exposed to subliminal messaging purchased 18.1 percent more Coke and 57.5 percent more popcorn. Vicary had found the holy grail of propagandists: implanting desire so deeply in the brain that it seems self-generated.

In 1957, Vance Packard and James Vicary separately announced that the splendid human mind had an elementary design vulnerability. If you had the right knowledge (cutting-edge behavioral science) and modern technology (mass media, tachistoscopes, airbrushes), you could control human minds on a massive scale.[7]

When James Vicary announced his experiment in "invisible advertising," public condemnation was violent and swift. Vicary's technique seemed to foretell an Orwellian world in which the very possibility of free thought was at risk. "This may well be," raged the *Washington Post*, "the most appalling assault upon the human brain and nerve system yet concocted by civilized man."[8] A different report called the technique "a rape of the mind."[9]

But the science touted by *The Hidden Persuaders* was no science at all. The whole field of unconscious messaging was built on the soupy (and loopy) foundation of 1950s-era Freudianism. Academic, corporate, and CIA researchers spent decades trying to make subliminal messaging work. In many scores of studies, they never could.[10]

And what about James Vicary's famous proof of concept? Less famously, he later confessed to fabricating his data as a publicity stunt. In other words, Vicary's method for saving his failing firm was simple con artistry. According to Madison Avenue lore, the con paid off to the tune of millions of dollars in contracts.

The Hidden Persuaders prophesied a brave new world where advertisers and apparatchiks would shape us like sculptors—but so subtly that we'd never feel the chisel. But there really is a scientifically validated method for stupefying the sentries of the conscious mind and smuggling ideas and yearnings directly into the unconscious. Ironically, James Vicary almost stumbled across it by accident.

It's possible—even likely—that Vicary never visited that New Jersey cinema and never superimposed any commands on William Holden's bare and brawny chest (the theater manager denied that he had). But as viewers watched Holden peacocking across the screen, ideas and values were still being slipped unobtrusively into their minds. The method of delivery was as ancient as humanity, even if its power is now validated by up-to-date science. This method of unconscious messaging goes by the humble name "storytelling."

By the 1990s, two-thirds of Americans believed subliminal advertising was both ubiquitous and wickedly effective.[11] It was neither. But though advertisers never actually went in big for subliminal messaging, they never gave up their dream of hidden persuasion.

Over the last decade, "storytelling" has likely surpassed even "innovation" and "disruption" as a top buzzword in the business world. A real and seemingly permanent culture of business storytelling has emerged, with the *New York Times* touting story's "almost irresistible power,"[12] with MBA programs introducing storytelling into their curricula, with companies employing Chief Storytelling Officers, and with the marketing guru Seth Godin emphasizing in bold print, "**Either you are going to tell stories that spread, or you will become irrelevant.**"[13] Business storytellers celebrate story because of its ability to deliver joy, values, and connection but also quite explicitly because it can act as a Trojan horse, smuggling messages into the defended citadel of the human mind.[14]

Thinking about stories as tools of mind control may seem ghoulish or sensationalistic. But it shouldn't. Flip back to the image of our Khoisan storyteller in Chapter 1. He has his hands up like a conductor and he's orchestrating all the images in his listeners' brains, the feelings in their hearts. He's brought his people into emotional attunement with each other, hormonal harmony, and neurological rhythm—all of which we could verify today in a proper lab.[15] The whole point for the Khoisan elder or any other storyteller is mind control. Powerful storytellers penetrate our skulls and take temporary control of our emotional and mental control panels—dictating the images that enter our minds and the feelings that swamp us. And they often do so with the clear intention of exerting short- or long-term sway.

This isn't to suggest that storytellers are often, much less necessarily, up to something sinister. Nor is it to suggest they should be blamed. There's no place of innocence in human communication. Humans are, by nature, intensely social primates who are always striving for position in our hierarchies. In the contest for sway, everyone is looking for an edge. So, yes, storytellers manipulate. But so does everyone else. For example, reason-based arguments are often built up with enormous deviousness using the whole armamentarium of rhetoric, which is the catchall name we've given to systems of linguistic and logical jujitsu that have been developed not so much to get at the truth as to get at sway.

The reason we should worry more about storytelling than other tools of sway isn't because storytellers are less moral but because they are generally more powerful. Storytellers enjoy a number of scientifically validated advantages over other types of messengers. First, and most basically, unlike some other forms of messaging, we love stories and the people who deliver them. Second, story is sticky (we process narrative much faster than other forms of communication and remember the information much better).[16] Third, stories rivet attention like nothing else (think about how little your mind wanders during your favorite TV show or a novel you can't put down). Fourth, good stories demand to be retold (think how hard it is *not* to spread that top-secret gossip or give away a spoiler), which means the messages in stories spread virally through social networks.[17] And all of these advantages are driven by the fifth and most important advantage stories have over other forms of communication: they generate powerful emotion.

Plato believed that the human mind was composed of three main centers. In a sound mind, the sphere of pure logic—Plato called it the *logistikon*—ruled over the baser centers of the emotions and the appetites. In an unsound mind, emotions and appetites flood the *logistikon* and anesthetize rationality.

Plato was hostile to storytelling because stories are all about catalyzing powerful emotional response. Stories *are for making us feel.* When we check the menu of options at our local movie theater, we don't decide merely on the basis of whether we like the star or the reviews are good. We're also choosing from a menu of emotions. If we want hilarity, we go for a comedy. If we want the thrill of fear, we see a horror flick. If we want a taste of righteous fury, we go for a revenge thriller. Sometimes we even like feeling sad, and so we fork over for a tearjerker.

Many writers have pointed out that storytellers must produce illusions of reality. But they've also pointed out that the illusion is highly artificial. Typical stories aren't like real life. They're like real life "with the dull bits cut out," as Alfred Hitchcock put it.[18] Put differently, they're like *real life with the unemotional bits cut out.*

Stories are for making us feel. But what are feelings for? When we say that we're "moved" by an emotional experience, we aren't speaking metaphorically. The English word *emotion* comes from the Latin *emovere*, "to move." This is why so many motion-related words in English—*mot*orcycle, *mot*ivate, loco*mot*ive, pro*mot*ion, de*mot*ion, and *mot*ion itself—share the Latin root, *mot.* To feel strong emotion is to be set in motion. Fear *moves* us to flee or hide. Fury *moves* us to fight. Regret *moves* us to apologize and mend our ways. Love *moves* us to protect and nurture.

Stories are all about emotion and, it turns out, emotion is the key element in human decision-making.[19] Rational argument is good mainly for preaching to the converted or disinterested. But it turns feeble when we need it most. Trying to reason people out of their emotional entrenchments can be worse than useless. When we try to reason someone out of deeply held views, we may produce so-called backfire effects that deepen resistance.[20]

Persuasion can generally make use of one of two cognitive channels: the rational or the dramatic. Communicators can rationalize by deploying arguments backed up with evidence, or they can dramatize with stories. Stories aren't favored in every situation. There are times when people want efficient delivery of unshaped information, and launching into a story will annoy not enthrall. But across studies carried out by different types of researchers, when it comes to persuasion, *dramatization usually beats rationalization.*[21] It's often not enough, in other words, for the audience to *get* the point. Sometimes for the better, often for the worse, we have to use stories to make them *feel* the point.

The Catch

But there's a catch, which I'll dramatize myself by describing the plot of Tommy Wiseau's infamous film *The Room* (2003).

Johnny is a rich guy who drives a Mercedes. But he's not stuffy or superior. Everyone loves Johnny, and his natural magnetism draws a community of loving friends into orbit around him. Even with his stressful work and caring for his stay-at-home fiancée, Lisa, he still finds time to play football and grabass with his friends up on the roof of his swanky apartment building. When the men get together, conversation invariably turns to the greatest of the great imponderables. *Women*, the men want to know, *what's up with them?* Sometimes, Johnny observes, women "are too smart." Other times they are "flat out stupid." And just as often "they are evil."

Thus, Wiseau announces his great theme: Women are odd little semi-humans who live according to an alien feminine logic and love playing games. Only, when compared to wholesome masculine pastimes like football and grabass, women's games are so wickedly devious that men can't hope to keep up.

Wiseau's film doesn't give us the kind of misogyny-lite of mansplaining around the office watercooler or manspreading on the subway. *The Room* gives us the pure, medieval thing. In what is, perhaps, the film's most infamous scene, Johnny's friend Mark tells a story about a promiscuous woman who's beaten near to death by a jealous boyfriend. Johnny erupts in joyous laughter and exclaims, "What a story, Mark!"

Johnny's fiancée, Lisa, is a lying, cheating personification of womankind's lustful and diabolical nature. By seducing Johnny's best friend, Lisa blows up their social world. When Johnny discovers the affair, he drives his friends away in despair. "Everybody betray me," he moans, "I fed up with this world." And then, after perpetrating a soulful, romantically lit sex act with Lisa's empty dress, Johnny shoots himself in the head.

The Room is best known not for its misogyny but for the hilarious incompetence of its writing, directing, and acting. Wiseau is among humanity's all-time worst writers (here's a fairly representative line of dialogue: "Promotion! Promotion! That's all I hear about. Heere [*sic*] is your coffee and English muffin and burn your mouth."). But on top of writing the barely coherent script, Wiseau also served as *The Room*'s director, producer, star actor, editor, and financier, thus demolishing his story at every conceivable level.

The Room is so far out on the spectrum of bad that it's mesmerizing and hilarious. It represents a uniquely failed attempt at genius that's as rare and instructive in its own way as actual genius. The result is a cult phenomenon with audiences showing up to shower the film with affectionate derision. The whole comic effect is based on the vast chasm between the soaring Wagnerian drama Wiseau was going for and the absurdist comedy he actually produced.

For *The Room*, Wiseau spent millions of his own dollars crafting a movie about a tragic human paradox: love is the answer to all of our problems, but women are unlovable demon-bitches. The message is very clear: the safest thing for men to do is avoid women altogether. But since this is clearly impossible, given how sexy and wily they are, men best be forever on guard.

But *The Room*, thank goodness, fails in its messaging. It's an object lesson in what storytellers must *not* do if they want their messages to actually sink in. I doubt it has persuaded men to be more wary of darling little succubae. And I'm sure it's never persuaded a single succubus to mend her ways.

This is because *The Room* is a consensus pick in rankings of the worst films ever made. And none of storytelling's messaging advantages kick in unless a story is actually good. What critics mean by "good" is soupy, subjective, and hard to define. But when ordinary people say a story is "good," they mostly mean that the story casts the spell of narrative transportation. In the language of the social sciences, all of the psychological, emotional, neurological, and behavioral effects of stories are "mediated" by transportation. That is, the more transporting the story, the better we like it, and the stronger its psychological effects across the board.[22]

So, the overriding question for a storyteller in search of sway is "How do I generate narrative transportation?" This isn't the place to dig deep into the tricks storytellers use to transport audiences (you can go to any good creative writing guide for that). But I do want to examine one key trick that connects to the "dark arts" theme of this chapter.

The Science of "Show, Don't Tell"

Once upon a time, Ernest Hemingway was at a restaurant with friends. He boasted boozily that his writing powers were so great that he could cram all the power of a novel into just six words. Hemingway's friends scoffed and bet him ten dollars each that he couldn't. The great novelist then scrawled six words on a napkin and passed it around the table. Each man blinked at the napkin for a moment before handing it along to his neighbor with a frown. Then they all fished out their wallets and gave Hemingway his sawbuck. Here's what was written on the napkin:

For sale: baby shoes, never worn.

Most people read this little story for the first time in brief confusion. *Wait. What?* And then there's a satisfying mental click. *Oh.* The tragedy unfolds in our minds: A hopeful couple—not rich—buys a pair of booties for their unborn child. There's a birth. And then a death. And then all the agony of exploded hopes.

Like some of the best Hemingway anecdotes, this one, alas, is probably untrue.[23] Scholars have shown that Hemingway almost certainly didn't compose this little story. But the idea that he did continues to spread as an all but undebunkable literary urban legend. Perhaps this is because the six-word story carries such powerful lessons that it seems like a great genius must be behind it. The tiny story shows how simple stories are, deep down. How little they require in the way of fancy words, complex thoughts, or basic originality in plotting. And it also illustrates how storytellers can confidently depend on audiences to do most of the work for them.

"Show, don't tell" reflects the collective wisdom of storytellers who've found that more is usually accomplished through subtle, indirect messaging than direct and explicit messaging. This is one of the tired clichés around storytelling that, like most of the other tired clichés, hold great validity. In recent years, researchers have found that stories with overt messaging are less persuasive than stories where the messages are implicit and indirect. As the communications scholar Michael Dahlstrom put it in the *Proceedings of the National Academy of Sciences*: "One of the few factors that has been found to hinder narrative persuasion is when the persuasive intent becomes obvious and audiences react against being manipulated."[24]

The Room's flaws, for example, are impossible to encapsulate short of a book-length treatise (if you'd like to read that book, see *The Disaster Artist* by Greg Sestero and Tom Bissell). But if I had to nominate just one thing that secures its place on the Mount Rushmore of bad art, it would be its insistence on telling, not showing. The film's dialogue carefully spells out, underlines, italicizes, and then repeats all of Wiseau's messages.

Now compare this to our literary urban legend, where all Papa had to do was scribble down six ambiguous words. We do the rest. In my mind, for example, I saw a couple placing the ad, not a single woman. And I imagined that the couple must be older and not rich. If they were young, they might have kept the shoes for another try. And if they were rich, why would they go through the bother of trying to sell them? Your mind probably didn't create exactly the same story as mine did. But, like mine, your mind probably took a bit of artistic license, going beyond what was written on the page.

Again, the six-word story isn't *telling* us anything about the parents, the baby, or the emotional aftermath of the tragedy. It's merely *showing* us something—an ad for baby shoes—and then we do all the work of constructing the meaning. The power of the six-word story is enhanced by the slight delay in comprehension that most people experience when first encountering the story. It stays with us (or we stay with it) until we can wrestle its meaning to the ground.

This ties into recent research on "retrospective reflection," which is the name psychologists have given to the final stage in the storytelling process. This is the moment when story consumers close the book, or walk out of the theater, and integrate ideas and information from the story into their preexisting worldview. Research suggests that stories are most persuasive when they're compelling enough, and open-ended enough, to pin us in storyland after the tale ends.[25]

To summarize, telling just *gives* us the meaning. Showing forces us to figure out the meaning for ourselves, and when we do this, we take ownership of that meaning. In this way, great storytellers play the psychological equivalent of the cuckoo bird's trick: they make us feel that the notions they've laid like eggs in our minds are actually our own.

"Secret Propagandists"

The wisdom of "show, don't tell" didn't evolve because it made for "good" art in some airy, aesthetic sense. It evolved because generations of tellers found, by trial and error, that this was (1) something audiences liked and (2) a potent way to concoct the drug of story and bring on the sway.

The fact is that persuasion is a contest, and it's often a little ugly. We'd like to think persuasion could be reliably accomplished simply by producing better and truer information. But persuasion isn't the same as instruction—as taking a blank slate and filling it up. Persuasion always begins with *dislodgment.* You have to move a mind from one place to another, which means overcoming inertia with some kind of force.

No one understands this better than the top-echelon storytellers. "In many ways," the great novelist and essayist Joan Didion says, "writing is the act of saying *I*, of imposing oneself upon other people, of saying listen to me, see it my way, change your mind. It's an aggressive, even a hostile act. You can disguise its qualifiers and tentative subjunctives, with ellipses and evasions—with the whole manner of intimating rather than claiming, of alluding rather than stating—but there's no getting around the fact that setting words on paper is the tactic of a *secret* bully, an invasion, an imposition of the writer's sensibility on the reader's most private space."[26]

The novelist John Gardner sees it the same way, defining storytellers as "*secret* propagandists"[27] and going on to say of stories that "nothing in the world has greater power to enslave."[28] I've called attention, by adding italics, to the shared key word in Didion's and

Gardner's quotes: *secret.* All of this bullying and propagandizing, to be most effective, should be sneaky and indirect.

As I've already noted, people in the business world love to illustrate the value of stories through the metaphor of the Trojan horse. It's a good metaphor because it aptly describes the sneaky essence of well-told stories and because everyone knows about the Trojan horse. But the Trojan horse isn't mentioned in Homer's story of the Trojan War (the *Iliad*), and it's only briefly referenced in his story of Odysseus's adventures after the war (the *Odyssey*). So, without an authoritative Homeric version of the tale, variants of the Trojan horse story proliferated in Greek and Roman literature.

One prominent version, told in Virgil's *Aeneid* and elsewhere, confronts the unlikely notion that the Trojans were simply duped by the big wooden horse. How stupid are they supposed to have been? They hated the Greeks and the Greeks hated them. Before the fall of their city made it proverbial, they already knew to distrust Greeks bearing gifts. So, in Virgil's account, the Greeks left behind a storyteller to shape the horse's meaning. This storyteller, Sinon, was the real Trojan horse. His mind contained the story that would crumble the Trojan walls.

In the tenth year of the war, Virgil tells us, the Greeks suddenly broke camp and sailed toward home in the black of night (in reality they just hid their boats behind an island not too far from the Trojan shore). Upon first discovering the massive horse in the deserted enemy camp, the Trojans also found the Greek soldier Sinon hiding in a marsh. At first the Trojans crowded around him in a violent fury, but they were soon lulled into placid gullibility by the power of his story.

Before the Greeks sailed away, Sinon explained, they decided that one of their number must be killed as an offering to the gods. To satisfy an old grudge, Odysseus conspired to make Sinon the victim. The other Greeks saw what Odysseus was up to but did nothing to stop him. However, before they could drag Sinon to the altar, the young soldier broke his bonds and escaped.

Pretending to be enraged and heartbroken by the betrayal of his countrymen, Sinon tells the Trojans his great lie. The Greeks built the wooden horse as an offering to Athena. To harm the horse would bring disaster to Troy. But if the sacred offering were brought inside the walls, the city would win divine favor. More than the Trojan horse itself, Sinon's story and his virtuosic way of telling it spelled Troy's doom. Without his story, the pretty horse would likely have gone down in ash, with all the Greek heroes trapped inside

The Trojan horse isn't a metaphor for story's power to do good. The Trojan horse is, after all, an engine of war. It holds a brutal holocaust in its belly—massacre, mass rape, slavery, cultural annihilation, infants hurled from the high towers. The Trojan horse is a metaphor for story's ready weaponization—something con artists, propagandists, fake news artists, cult leaders, hucksters, conspiracy merchants, and demagogues understand in their bones. In place of the once-proud city, the Greeks leave a smoldering ruin. They win the war but make monsters of themselves.

Storytelling's Forever Problem

Let me tell you about a wicked experiment involving a boy named Stefan who's an all-around psychological and emotional wreck but a prodigy when it comes to writing computer code. At just sixteen, Stefan catches the attention of a video game design firm that commissions him to make a game based on his favorite choose-your-own-adventure novel.

The disturbed young man goes to work, creating a game that crams the space of narrative possibility with branches that twist out and twigs that curl back in. Stefan stops sleeping. He barely eats. The blearier and more malnourished he becomes, the more his paranoia grows. Actual life, he begins to sense, is exactly like a choose-your-own-adventure story, where the future branches out into a multitude of alternative realities. Every choice we make, every word we say, rewrites the future in big ways or small. But what if we aren't the ones doing the choosing? Stefan becomes convinced that some dark intelligence is making all of his choices for him, "playing" him as though he himself were a character in a game.

Sometimes, Stefan believes, this cruel intelligence forces him to commit homicide, sometimes suicide. Sometimes Stefan is forced into choices that lead him to live happily ever after and sometimes he ends his life rotting in jail. And always, just as the plot of his life climaxes, it loops back to the beginning, branching and bending in new and disturbing ways. Down one narrative path, Stefan discovers a terrible truth: he's just a film actor—Fionn Whitehead—acting his way through a preordained script. All along, Stefan learns, he's only been *pretending* to be distressed, only *pretending* to kill or die. But if he's forced down a different path, Stefan will discover that he's not an actor after all. Neither is he a schizophrenic or an abuser of hallucinogens. He's actually a lab rat in a mind-control experiment run by mad scientists.

I'm describing the tangled plots of *Bandersnatch*—a spinoff of the sensational Netflix series *Black Mirror*. *Bandersnatch* offers a fascinating violation of the fourth wall—the imaginary barrier that separates the pretend world of actors from the real world of the audience. Violations of the fourth wall typically involve actors breaking offstage (or out of character) to interact with the audience, thus dissolving the line between the pretend and the real. But in *Bandersnatch* it's the viewers who break into the fiction world to accost the main character and steer his fate. At a crucial point in the *Bandersnatch* experience, we're given the option of letting Stefan in on the ugly truth. We can contact him through his computer and inform him, whether out of cruelty or kindness, "I am watching you on Netflix. I make decisions for you."

But what if the actual target of *Bandersnatch*'s mad-science experiment isn't Stefan but Netflix customers sitting on their couches and keying decisions into their remotes that reveal personality traits they'd rather keep hidden? For example, will viewers succumb to a sadistic impulse and compel Stefan to commit the unprovoked murder of his loving but flawed father? Will they then have Stefan butcher daddy's corpse in the bathtub in hopes of hiding the evidence? Will they force him into a highly unchivalrous fistfight with his female psychotherapist? Will they stand the frail boy on a high-rise balcony and shove him over with a button click?

Well, yes, in the case of the present viewer, "they" would do all of these things. *Bandersnatch* is akin to famous experiments probing our capacities for cruelty—the Stanford prison experiment or the Milgram experiment with simulated electroshock. How cruelly will we behave, how much real-seeming suffering will we inflict, if we think it's "allowed" and no one is judging?

Like scientists ticking boxes as a rat runs a maze, data scientists at Netflix watch carefully as we puzzle our way through *Bandersnatch*'s labyrinthine plots. Based on our clicks, Netflix can infer a lot about us: whether we prefer crazed, mindless action scenarios or slower, more cerebral plots; whether we prefer romance to humor, unlikely plot twists to more realistic outcomes. Based on the breakfast we choose for Stefan, they can even infer whether we prefer Frosted Flakes to Sugar Puffs.

Netflix admits that it collects data on *Bandersnatch* viewers in order to "inform the personalized recommendations you see in future visits" as well as to help "determine how to improve [*Bandersnatch*'s] model of storytelling."[29] Put differently, when we stream *Bandersnatch*, we become subjects of an experiment in which Netflix harvests psychologically revealing information in order to exert control over our minds and behaviors. At a minimum, Netflix will use this data to predict our preferences, to maximize our time on their platform, and ultimately to drive the behavior that matters most to them—making sure we keep paying dues.

Netflix must grapple with a "forever" problem in storytelling. Successful stories all around the world, and as far back as we can read in history, share strong regularities that have been the subject of scrutiny since at least the time of Plato and Aristotle (this topic is covered in Chapter 4). But it's just as obvious that not all stories please all readers equally. Implicit in all generalizations about what works in storytelling is an average story consumer upon whom a story works or doesn't. But this average story consumer is a statistical abstraction, useful for big-picture thinking about stories, but never to be encountered in the real world.

Actual flesh-and-blood storytellers address themselves to actual flesh-and-blood people, who vary in a multitude of ways. For example, researchers have studied how basic demographics like race, class, gender, educational level, and age—as well as basic personality traits like openness to experience, empathy-proneness, and social intelligence—affect responses to stories.[30] In some ways, this research just confirms the obvious: some people (me) might get teary-eyed during a sappy commercial; others watch *Schindler's List* dry-eyed. You might prefer tales of true crime, and I might have an unwholesome appetite for voyeuristic biographies about the chaotic lives of great artists.

But in other ways, the research isn't as obvious as it may seem. For example, interesting differences emerge along gender lines.[31] Researchers find that women's higher average empathy, social intelligence, language skills, and fantasizing ability (yes, this can be measured) make them more transportable than men on average.[32] This would suggest that anyone who isn't making stories with women in mind is missing the ideal audience—which makes creating more opportunities for female producers, writers, actors, and directors not just good ethics but good business.

Netflix and other digital storytelling platforms—including online newspapers, magazines, and social media sites—collect a lot of data. They know what type of stories we like. They know when we get a little bored and take a bathroom break—or get so very bored that we just give up and don't come back. And they use this information to inform two critical judgments. First, which stories perform best for consumers in aggregate? Second, which work best for each individual so that Netflix can steer just the right shows to just the right people?

Perhaps it's hard to feel worried about this. After all, what's so bad about Netflix nudging us toward stories we're likely to love? But story-targeting technology can be used toward darker ends.

STORyNET

Almost a decade ago, the Defense Advanced Research Projects Agency, better known as DARPA, held a conference on storytelling. DARPA, a semisecret agency within the US Department of Defense tasked with developing new military technologies, is sometimes referred to as "the heart of the military-industrial complex."[33] This conference brought together neuroscientists, computer scientists, psychologists, and storytelling experts from the humanities for a new funding initiative on "Stories, Neuroscience, and Experimental Technologies."

In an apparent effort to make the program seem just as dystopian as it may in fact have been, they didn't call it SKYNET, which was already taken by the *Terminator* movies, but STORyNET.[34] DARPA administrators were up on new science suggesting that specifically story-based forms of information warfare had special potential. At the time, the Pentagon was deeply embroiled in wars in Iraq and Afghanistan. Al-Qaeda and other adversaries were not only holding their own on the battlefield but also winning the war of rival narratives on the internet. DARPA wanted to take stories apart, study their most powerful effects, and then put that knowledge to work through technologies that would multiply their persuasive power.

Traditional war propaganda—think radio broadcasts or leaflets dropped from the belly of a bomber—is the messaging equivalent of a dumb bomb. It's indiscriminate. It hits everyone with the same one-size-does-not-fit-all message. DARPA imagined a new type of story that would, like the shapeshifters of speculative fiction, constantly adjust itself to the unique psychology of every individual story consumer. In short, these stories would be like bespoke suits, sharing certain stable features with all other suits (e.g., having four tubes for arms and legs), but also being perfectly and precisely tailored to the specific neuroanatomy of each individual.

The idea of your TV reading your mind and adapting stories to both the stable features of your personality and the quickly shifting features of your mood might seem awfully far-fetched. But as we saw in Chapter 1, our subjective reactions to stories correlate reliably with measurable physiological reactions. If a story scares us, for example, we sweat a little more, we breathe harder, our hair stands on end, and our pupils dilate. You don't have to crack the skull to observe the brain on story. You just have to collect physiological data that correlate in a probabilistic way with known brain states.

A DARPA-funded research team led by the neuroscientists Jorge Barraza and Paul Zak adapted a true story about a terminally ill little boy and his grieving father, which included an appeal to donate to a charity serving sick children. Using basic polygraph-type devices that captured only cardiac information and sweat rates, they wanted to see if they could predict whether given audience members would "buy" the message. By eliminating people whose responses showed low emotional arousal, they were able to *predict with 80 percent accuracy* who would eventually give to the charity. In other words, they didn't just read the viewers' minds. They did something more impressive: they read the future.[35]

But STORyNET had an even more ambitious goal. In a choose-your-own-adventure story like *Bandersnatch*, you make frequent decisions that cause the narrative to veer down different paths. But what if a computer could read your mind microsecond by microsecond and push you down different narrative paths based on an array of physiological indicators known to correlate reliably with your mental and emotional state? This data could be scraped up—as it's already being scraped up in China—from an array of web-connected cameras, microphones, TVs, and other devices.[36] And, finally, what if the computer did so not to maximize the viewer's pleasure so much as his or her *compliance* with the story's message—whether the message was an uplifting one about sectarian harmony or a sales pitch or a sinister narrative about a minority group?

Stealth choose-your-own-adventure stories, uniquely tailored to maximize message impact on individual audience members, are technologically feasible. If they fail to materialize, it won't be because of technical barriers. It will be because a cheaper and dirtier form of story-based manipulation has already beaten this one to market.

The New Panopticon

There are two ways of creating bespoke stories with maximum manipulative force. The first is DARPA's effort to use a complex technological array to sweep a story consumer for data. Then, based on what's learned, the consumer would be pushed down the branching tree of narrative possibility in a way known to achieve message compliance.

But governments and corporations are already perfecting a second way of doing this. Instead of learning about people on the fly, as they respond to a story, it's easier to learn everything about them in advance—all their demographic data, their quirks of personality and politics, and even the sexual urges that ravage their imaginations but have never been confessed to a single soul. Then, once everything in a person's identity has been reduced to a collection of bar graphs and scatterplots, that person can be exposed to narratives known to be maximally effective to people who share similar psychometric profiles.

This requires powerful technologies of surveillance. In the eighteenth century, the social reformer Jeremy Bentham arrived at an idea for a perfect surveillance technology called a panopticon, a type of prison that would be laid out on a hub-and-spoke design.[37] The jailer, positioned in the hub, could see everything that the inmates were up to. But here's the really clever part: inmates, thanks to a system of lighting and blinds, couldn't see back into the hub. So, even though the inmates know the jailer might not be watching, they must behave as though he always is.

The word *panopticon* is from the Greek meaning "all-seeing," and it's a pretty good description of the digital surveillance we're all living with now. The most important difference between Bentham's panopticon and modern digital versions is that Bentham *wanted* people to feel as though they were being watched even when they weren't (the ubiquitous two-way screens in *1984* serve the same purpose). But the new digital panopticon is designed to make us feel that we *aren't* being surveilled, even though we always are.

Moreover, while Bentham's panopticon could only provide visual information, the new digital panopticon reads minds. As we move through our days, we shed data about ourselves like dandruff. The "free" digital economy, as many technologists have observed, is one where the user's information, thoughts, aspirations, and precious attention are the actual products up for sale.

What does the digital panopticon know about me, a relatively modest tech user with a light social media footprint? Well, everything. My phone knows where I am at all times and can even predict where I'm going before I leave. Thanks to a nutrition app I use, it knows what I eat in detail and could probably make excellent guesses about my mood based on when the binges occur. Thanks to other apps, it knows when I go to sleep at night, when I wake up, and how I'm doing with my exercise routines. It knows, through my Google searches and other web traffic, *exactly* what's on my mind. Thanks to Siri, it even knows what I sound like when I'm in a good mood or a bad one. In total, even if this information is cut up and spread around, companies like Amazon and Google know far more about me than any flesh-and-blood human I can name. They know, and can pretty accurately predict, my taste in books, music, movies, and journalism. They know my politics. They know my hobbies—the ones I stick with and the ones I let slide. They know my secret yearnings and shameful vanities. They know things about me that I'd never be reckless enough to write down in a diary.

The digital panopticon can see what I'm *really* like based on how I actually behave and think minute by minute. It sees my real face, not the different masks I show the world. Even I don't know myself with this kind of intimacy or precision. The panopticon sees me without the self-serving egoistic biases that deform human self-understanding. And it remembers everything it has ever learned about me, whereas I forget almost everything.

The cold scientific vocabulary of surveillance capitalism seems far away from the warm old craft of tale-telling. But these "behavior modification empires," as the computer scientist Jaron Lanier calls them, collect this data largely toward the goal of targeting us with narratives that are more engaging, more emotionally arousing, and ultimately more persuasive.[38] In the end, it's about getting us to buy whatever's on sale—from a harmless new widget to a socially toxic virus of thought.

Blitzkrieg 2016

On May 21, 2016, two groups of protesters descended on the Islamic Da'wah Center of Houston, Texas. One group comprised members of a Facebook group called "The Heart of Texas," which was devoted to celebrating Texas heritage and to political issues such as gun rights and immigration control. The other group of protesters had been mustered by a different Facebook group, the United Muslims of America, and they advocated for causes like immigrant rights and gun control. That day, Heart of Texas members were responding to a call to gather at the "shrine of hate" (aka the Islamic Center) to "Stop Islamization of Texas." Members were told to "Feel free to bring along your firearms, concealed or not!" and many did. When news of the protest spread, members of United Muslims of America rallied to protest the protesters and make a statement about the scourge of Islamophobia. The two groups of protesters waved signs for a while, shouting at each other over a cordon of cops, until they all got tired and went home.[39]

Just another day in America.

Only it wasn't. None of the protesters knew that they were just puppets dancing on invisible wires, controlled by a puppeteer from thousands of miles away. Instead of using wires to bring the two crowds together for a war dance, the puppeteer only had to tell them different stories.

The Heart of Texas and the United Muslims of America were two of almost five hundred Facebook groups created by the Internet Research Agency (IRA) in Saint Petersburg, Russia. The IRA is the now famous "troll farm" at the center of the Russian influence machine that was unleashed on the 2016 American presidential election. Both Facebook groups were highly successful propaganda efforts. The Heart of Texas accumulated 250,000 members, and United Muslims boasted 300,000. In total, posts to these two sock puppet groups were liked more than eight million times and shared more than ten million times.[40]

It's misleading to call the IRA a troll farm.[41] It suggests a building full of pimply, adolescent shitposters—the kinds of losers who have nothing better to do than befoul the comments on YouTube videos with cruel, profane posts. No, the IRA is a story factory. Since 2013, it has trained thousands of Russians, often with backgrounds in journalism or PR, to serve as shock troops in a story war. These propagandists are trained to see the fault lines in a society and then drive wedges into them with enormous wallops of power.

The Heart of Texas members were told stories in which the whole way of life of good, native-born Texans was under threat from foreign infiltration. The United Muslims were fed stories that celebrated Muslim American identity and the most clear and present threat to it: the types of deep-red Americans represented by the Heart of Texas. Sometimes the stories were pure fake news. Sometimes the stories were completely true, but carefully preselected to inflame the resentments of one community or the other. Both sides were made to see themselves as the good guys in the story—as cornered protagonists who have to fight their way out.

The Russians saw that Facebook was the greatest mind-control machine in history. With the right kind of data, they could turn the machine not into the empathy and connection machine it bills itself as but into a propaganda machine that churns out hostility and division.

This effort is usually referred to blandly and vaguely as "the Russian influence campaign." Better to say what it actually was. It was a blitzkrieg of stories. It was a weaponization of stories and of people's built-in vulnerabilities to story. The short-term goal was to tip the election toward the Republican candidate by distributing narratives that would rev up Republican voters while demoralizing Democrats. The long-term goal of the Russian effort was to do lasting damage to America by inflaming all of our tribal resentments.

The Russian blitzkrieg of narratives is among the most brilliant, devastating, and far-reaching propaganda attacks in history. It's still not clear whether this effort to swing the election was decisive or not, in a contest that was decided by a few tens of thousands of votes in a handful of purple states. But they certainly achieved their bigger goal of weakening and fracturing the American colossus—of so obsessing us with our internal squabbles that we can't maintain our position as a stabilizing force on the world stage, much less extract a high enough price to deter further attacks. The Russian intelligence services used all sorts of weaponry, like memes and infographics and fake news stories. But what these all had in common was the intent of creating rival narratives that would clash, and spark, and ultimately turn the country against itself like a dog spinning round and round, getting ever dizzier and weaker as it strains to bite its own ass.

The Poetical Philosopher

If there's one thing most people know about *The Republic* it's "The Allegory of the Cave," which is still one of the most widely taught texts in college courses.[42] If the average person knows a second thing about *The Republic*, it's probably the notion I presented to open this chapter: Plato was so monomaniacally opposed to stories that he hoped to toss all the storytellers over the city wall.

But wait! The two facts are glaringly incompatible. "The Allegory of the Cave" *is* a story (allegory is a literary genre). And Plato's protagonist, Socrates, while based on a real guy, isn't just the roundest, quirkiest character in all of ancient literature, he is himself a talented and indefatigable storyteller. The whole *Republic* is, in fact, a first-person story told by Socrates himself. He begins by explaining how some of his rich buddies waylaid him and, with mock threats of violence, forced him to come over and talk philosophy. And inside this conversation are many little stories and small comic asides and large mythical masterpieces of Plato's own invention. This all amply justifies Sir Philip Sidney's claim, in *The Defence of Poesy* (1595), that "of all philosophers" Plato is "the most poetical."[43]

So *The Republic* shows us Plato using all of his story craft toward an attack on the craft of story. He's telling a tale that would end—if the moral of the story were taken half-seriously—with getting him tossed from his own utopia.

What's going on here?

One possibility is that Plato doesn't see the irony of writing an anti-poetry poem. Maybe he doesn't notice that he's condemning fiction through fiction—which would make *The Republic* a self-devouring ouroboros of a book. But this seems unlikely. What's more likely is that those many writers who say Plato wants to banish the poets have missed his point (or failed to read far enough into his book!). No one has understood story's dangers better than Plato—which is why his first impulse was indeed to banish all the storytellers. But neither has anyone more fully appreciated story's society-building potential—which is why he ultimately decided not to.

Twenty-four hundred years ago, Plato reached the same conclusion I have: story is the main tool for controlling and shaping human behavior on a massive scale. Far from being against stories in a practical sense, there has never been a great thinker so alive to the power of story and so fully committed to its use.

3

THE GREAT WAR FOR STORYLAND

HERE'S THE GREAT PROBLEM ADDRESSED BY PLATO'S *REPUBLIC*: How can human beings live by reason instead of dying by unreason? Plato never tries to hide the fact that his answers to the question are as radical as can be imagined. First, Plato would run just the *free* poets out of town—the ones who refuse to channel their talents toward the needs of the state. Second, he'd ban the most emotionally arousing styles of tale-telling and put what's left under the jurisdiction of his philosopher-king. Much of what was already on the book shelves would be trashed. But other works would need to be censored, revised, and repurposed. The main target would be Homer's *Iliad* and *Odyssey*, the closest thing the Greeks had to a Bible. Long stretches of *The Republic* are given over to describing a cretinous book-mutilation party in which everything Plato objects to in the epics—which is most everything—would be either tweaked, overhauled, or deleted until we end up with an *Iliad* and *Odyssey* made sanitary and proper.

Plato also calls for a system of communism pushed to its logical limits. He would abolish all private property and put everyone—from ditch diggers to doctors—at the same economic level. Because wives in Athens were basically classified as a husband's property, wives would be held in common, too, with offspring raised collectively (authorities would also intercede to arrange matches for eugenic purposes).[1] Plato's system would seek, then, to establish a communism of affection in which traditional romantic love would be abolished as would love between parents and children.

In short, Plato suggests that the only way to solve humanity's gravest problems is to annihilate love—of beautiful poems, of children and siblings, of wives and husbands. To actually make such a dream come true, Plato saw, there's not enough force in the world. To enact his perfect vision, he would therefore need to win compliance from people who are too oafish to be moved by pure logic. And how could he possibly achieve this when the steps toward perfection run so painfully against the grain of human nature?

Stories. Lots of stories.

Far from being a kingdom stripped of stories, Plato's *Kallipolis* is a kingdom in which stories are baked into the very bricks. The only difference is that the stories don't arise organically but are engineered by a class of philosophical supermen who have, through a combination of eugenic breeding and specialized training, been perfected for the task.

In the end, the ruler of Plato's ideal republic would be a storyteller-king as much as a philosopher-king. Plato sets out to describe a dictatorship of reason and ends up describing a dictatorship of fiction in which every tale—from the smallest nursery rhymes mothers sing to their children at bedtime to the grandest myths of human origins and endings—would come under total state control. It's through his monopoly on propagandistic stories and myths that Plato's king would rule.

Karl Popper was one of the greatest philosophers of the twentieth century. And he made his bones by going after Plato with a cudgel. His thrillingly iconoclastic masterpiece, *The Open Society and*

Its Enemies (1945), unfolds like a gory scene of philosopher-on-philosopher crime. In it, he bashes Plato's *Republic* as an evil how-to manual for building a totalitarian dystopia.

I think he's too hard on Plato. But Popper's right that rule number one of every totalitarian utopia dreamed up from Plato's time forward is *control and monopolize the story.* This applies uniformly to all the nightmarish totalitarian experiments of the twentieth century, including those of Nazi Germany, the USSR, North Korea, Cambodia under the Khmer Rouge, and both Maoist and modern China.[2] All of these regimes insisted on keeping a monopoly on all forms of storytelling, from journalism to art. Storytellers who wouldn't bend to the messaging needs of the state were killed or banished to a gulag, as were those who refused to at least pretend to believe in fictionalized versions of reality. These regimes understood that the state can't dominate the real world unless it first dominates storyland.

Take the Catholic Church, which is ruled by a classic storyteller-king. Only they call him pope, not king. But he lives in a palace, sits on a throne, is bedecked in regal finery, and wears a crown on special occasions. And he's surrounded by similarly bejeweled and bedecked cardinals known as "princes of the church," and all others must bow to kiss his ring. And all of his power and perquisites are based on his claim to know the deepest, truest version of the Christian story and to tell it as infallibly as God himself.

For more than a millennium, up to the Protestant Reformation of the sixteenth century, many popes worked to establish a papal theocracy that would overspread the planet and reduce the world's princes and potentates to vassalage. At the height of their power, the Papal States came as close as any entity ever has to proving that "the storyteller rules the world." Through their control of the Christian story, the church exerted enormous sway over the formerly disconnected and insular European peoples and wove them into a transnational civilization known as Christendom. The church had ultimate authority on all spiritual matters but also enormous worldly muscle—the muscle to control nations, command armies, coerce kings, and kill dissenters and apostates.

The Catholic Church told a story about ultimate reality and wrote its own enormous power into the tale, laying down a toll road to heaven that ran through the treasure rooms of the Vatican. To ensure the story would be theirs alone to tell, the church gave masses in Latin and made it a capital crime to translate the Bible into the common European tongues. So long as all holy matters were communicated in dead tongues, which only churchmen and educated aristocrats could understand, common people couldn't question whether the rules laid down by priests actually matched up with the scriptures.

Heretics would pop up now and again to challenge the story, and with it the church's power, but they were easy to dispose of through creative expressions of ultraviolence like live burning, live disembowelment, live dismemberment, or some combination of all three. These torture-murders were carried out in public for their entertainment value, yes, but also to advertise the price of resistance. Eventually, in prosecuting their story war against the Protestants, the church invented the world's first ministry of propaganda, the *Congregatio de Propaganda Fide* (*Congregation for the Propagation of the Faith*), for the purpose of spreading Catholicism in non-Catholic lands.

In all of this, the popes exercised the kind of ruthless control over narrative that Plato could only dream of and that other storytelling kings, like Lenin, Hitler, Mao Zedong, and Kim Il-sung, tried to establish through their own combinations of myths, propaganda ministries, and terroristic violence against heretics.

No institution has ever enjoyed such a powerful storytelling monopoly over such an extended period of time, even if the centuries since the Protestant Reformation have seen a steady erosion of the church's hard power. Today, the pope remains the absolute monarch only of the bling-crusted Vatican complex in Rome. But through his claim of ultimate authority over the Christian story, the pope still wields enormous soft power—enormous sway—over the world's 1.3 billion Catholics.

Art Is an Infection

This is a book about a war. It's the most important war in the history of the world. It goes back to the origins of humanity and it can never end. But the great war for storyland is nonetheless always contested with great ferocity because, as the saying goes, the storyteller rules the world. The grander the narrative, the more awesome the potential sway.

The word *story* carries connotations of levity, pleasure, and harmlessness. So the idea of a "story war" sounds almost like a contradiction in terms, like imagining two rainbows trying to devour each other. But story war is arguably the most pervasive and consequential form of human competition: the struggle to impose one's narrative at the expense of the other guy's, whether in geopolitical or marital power struggles.

Humans fight by clubbing each other with fists or insults, by piercing each other with flying arrows or bullets, by vaporizing each other with explosives, and by sickening each other with chemicals. But the most significant conflict, if also the most overlooked, is the forever war of stories over the terrain of the human imagination.

The world is spinning down a post-truth vortex of media bubbles, fake news, and feral confirmation bias. The more consensus reality dissolves, the more we're living in a de facto storyland, and the more the future will be shaped less by the facts than by a war of rival storytellers.

This all raises a critical question: How do you win a story war?

What works in the war of stories is similar to what works in what Darwin called "the war of nature": not what's superior in any moral or aesthetic sense, but whatever happens to propagate. A great place to begin exploring why some tales spread and some don't is in the humble folktale. Folktales are stories that rise up and circulate among a society's common folk and are hard to trace back to an original author. This makes them a perfect natural laboratory for figuring out what works—or doesn't—in a free market of story.

It's wrong to think of folktales as a genre of the past not the present. Today, tales spring up from the folk as they always have, but now they spread with incredible speed and ease on the internet. The classic jokes we tell, for example, are humorous folktales that have propagated through society simply because they're strong stories—narrative grenades that cram big wallops in little packages.

Or think of the classic urban legends about men waking up with no kidneys after partying in Las Vegas or "Little Mikey," the Life cereal mascot, dying in a horrible explosion of cuteness after consuming Pop Rocks candy and soda. We know these tales because they triumphed over legions of potential competitors that never spread. Even prior to the internet age, urban legends like these traveled the globe by word of mouth. Far from being amplified by authorities in the mass media, they spread *despite* top-down efforts to snuff them out.

Leo Tolstoy is best remembered as the author of great fiction like *War and Peace* and *Anna Karenina*, but he was also a daring philosophical thinker, and in 1897 he published a book called *What Is Art?* He wanted to pin down the essence of all art, but most particularly his own form of art—storytelling. And he came up with the best concise definition of art that I know of.

Tolstoy defines art—including story art—as emotional infection.[3] Good art infects the audience with the artist's emotion and ideas. The better the art, the stronger the infection. The better job it will do of working around whatever immunities we possess and planting the virus. Tolstoy reached this conclusion through artistic intuition, not science, but more than a century after his death, this is what psychologists are finding in the lab.

As we have already seen, emotion drives story's persuasive power. Strong stories usually generate strong emotions, and emotion acts as a solvent on our skepticism. But emotion also predicts whether stories are shared.[4] In fact, the strongest predictor of virality is, as Tolstoy predicted, the emotional punch the message packs. As the psychologists Robin Nabi and Melanie Green explain, our impulse to share emotional stories "has been widely documented across cultures, gender, and age groups. In fact, the more intense the emotional experience or the greater the emotional disruption, the more likely it is to be socially shared, and shared repetitively over an extended period of time."[5]

For example, the more an urban legend instigates the emotion of disgust, the more it will be remembered and passed along. Studies by the psychologist Joseph Stubbersfield show that "stories involving deep-fried rats, pus-filled tumors in chicken burgers or accidental incest are more likely to be culturally successful than tamer, less disgusting, ones."[6] And a study of politically oriented tweets showed that for every word researchers classified as emotional, Twitter users were 20 percent more likely to retweet the message.[7]

But when it comes to predicting virality, all emotions are definitely not created equal. Emotions can be divided into two types: *activating emotions* such as rage, anxiety, and elation, and *deactivating emotions* such as contentedness and despair. Activating emotions are physically arousing—boosting heart rate, respiration, and blood pressure as we gear up for action. Deactivating emotions are physically enervating; they incline us toward inaction. It's the difference between the rage that makes us want to leap out of bed and fight and the despair that makes us want to crawl under the covers and hide.[8] Stories that evoke activating emotions inspire us to retell them; stories that evoke deactivating emotions cause us to clam up.

Let's see how these findings apply to what may be the most reliably contagious collection of stories ever told—the gospels of Jesus.

The Storyteller-King of Kings

Around the year 30, Jesus of Nazareth, one of many charismatic rabbis preaching the end times, was killed by crucifixion and entombed. Three days later, some of his followers began telling a miraculous tale. Yes, the rabbi had died in the most gruesome and humiliating way, but he'd since risen to life. Jesus was not merely the Messiah of the Jewish people; Jesus was God himself. At first, almost no one believed them.

At the time, there were about twenty Christians on earth—Jesus's male disciples and a handful of women. They were all powerless nobodies. Jesus himself had been a workingman from a rural backwater, and all of his followers were poor and unschooled. Moreover, as Jews, they were all members of a weak and oppressed tribe within the mighty Roman Empire. And yet somehow, in a few short centuries, and in spite of sometimes-vicious suppression, this movement of powerless people culminated in the most explosive religious revolution the world has ever seen.[9]

At first, the handful of Christians were mocked as a ludicrous cult. Three hundred and thirty years later, Christianity had tens of millions of followers and was named the official religion of the Roman Empire. In the centuries that followed, the story of Christ would come to dominate every aspect of Western culture (art, philosophy, law, sexual norms, and so on), eventually becoming what it is today: the most successful religion on earth and the most powerful kingdom of storytelling in history.[10]

God isn't dead or even ailing. Even in our scientific age, vast majorities of people continue to believe in some form of divinity. But individual *gods* can die, even awesomely powerful ones. Gods die when believers stop uttering their names, stop pouring their libations, stop telling their stories. The once-invincible gods of Greece and Rome—Jupiter, Mars, Venus, Neptune, Juno, and so on—are dead as rocks. And it was Jesus Christ, the pacifist God, who killed them. He starved them of their sacrifices of blood and smoke until they withered down to what they are today: mere literary characters trapped in old books.

The early Christian evangelists were making a colossal "ask." They were asking pagans to reject the gods of their fathers—gods who'd been worshipped for thousands of years, and whose strength and vindictiveness few had previously doubted. They were asking them to leave their faith communities and alienate their friends. They were asking them to denounce their old gods as fictions or demons and to put all their faith in an upstart god from the nowheresville of Israel. If they did so wrongly, there could literally be hell to pay.

How did this single peace-loving god, worshipped by a handful of zealots, rout the battalion of ancient gods who dwelled on Mount Olympus—the gods of the high and mighty Caesars? For the Christian faithful, the answer is simple. Christ threw open pagan hearts. But why, then, didn't he throw open Jewish hearts (relatively few Jews converted) or those of the 68 percent of human beings around the world today who don't identify as Christian?[11]

In his influential book *The Rise of Christianity*, the historian of religion Rodney Stark lays out a sweeping explanation based on the combined effects of higher Christian birth rates and the influence of vast and slaughterous epidemics. Christianity happened to come on the scene, Stark argues, at a time when the Roman Empire was ravaged by plagues that killed millions of people and crudely pruned every family tree. Such calamities can shake a people's faith in the power and goodness of their familiar gods, leaving them vulnerable to poaching by an ambitious competitor god promising love, salvation, and joyful reunions in the afterlife.

So, according to Stark, the Christian God got lucky. He was in the right place at the right time, just when the pagan confidence in the old gods was waning. But the biblical historian Bart Ehrman contends that Christianity also had stark advantages as a collection of interrelated stories. Christianity didn't spread in any top-down fashion. Unlike the muscular church of later centuries, the early church was far too scrawny to coerce people into belief, to scourge and burn infidels, or to spread belief "by the sword."

Christianity spread through organic word-of-mouth storytelling. Ehrman describes the humble way the "good news" began to take over the Western world:

> A Christian woman talks about her newfound faith to a close friend. She tells the stories she has heard, stories about Jesus and about his followers. She also tells stories about her own life, and how she has been helped after prayer to the Christian God. After a while this other person expresses genuine interest. When she does so, that opens up more possibilities of sharing the "good news," because she too has friends.[12]

In short, Ehrman describes the triumph of Christianity as a triumph of viral storytelling.[13] Christianity's story was superior in the only sense that matters here: those who heard the story were more likely not just to believe it but also to retell it.

But why?

The Christian religion is made out of powerful stories. Christ himself was a practiced storyteller, traveling around telling the instructional tales we call parables. The first evangelists and gospel writers followed his lead, spreading stories from Christ's life for the purpose of gaining converts and saving souls. But Jesus was far from the first holy man to teach through parables—even if he was among the best. All religions are made out of stories, and all religions spread according to how often and how ardently those stories are retold.

How did the Christian stories, with their goody-two-shoes ethic of "love thine enemy," outcompete the established pagan myths, which, with all their sex and violence and soap-operatic twists, might seem a lot more compelling?

As Ehrman explains in *The Triumph of Christianity*, two uncontroversial advantages were woven directly into the Christian story. First, unlike Judaism or the paganism of Greece and Rome, Christianity is a missionary religion. Once a person receives the good news, it becomes his or her holy obligation to pass the stories along. To keep such good news to oneself would be greedy and sinful—an act worthy of punishment. This obligation to spread the gospel resembles the coding built into chain letters: "Share this letter with six people or something terrible will happen."

Second, Christianity was, like Judaism, a fantastically intolerant religion. Roman religion was polytheistic and highly tolerant of competitor gods. Jupiter didn't much care if you also worshipped Mars or even Jesus so long as you didn't stint on *his* offerings of blood and wine. The Judeo-Christian God, on the other hand, is jealous and tolerates no rivals. (The First Commandment: "I am the Lord thy God, thou shall not have any gods before me.")

So, the Christian stories came with two bits of coding—evangelicalism and monotheistic intolerance—that encouraged believers to pass the story along while at the same time immunizing their minds against infection by any future upstart gods.

But Christianity's improbable success in the story wars can also be explained, at least in part, by the above-mentioned connection between a story's intrinsic emotionality and its virality. In its contrasting visions of heaven and hell, Christianity offered much sweeter carrots and much bigger sticks than the pagan gods could muster. In Greek and Roman mythology, the vision of the afterlife was hazy and inconsistent, with different myths suggesting different outcomes. But on the whole, the vision was drab and bland. For example, Homer's version of Hades is like a giant waiting room where souls stand around together for eternity just being bored—missing their bodies, missing the sunlight, missing the savor of wine and meat. Except for a small number of great men who were sent to the paradise of the Elysian Fields, good people and bad people were sent to the same joyless waiting room. The religion did offer incentives for good and bad behavior, but they were mainly terrestrial: If you sinned, the gods might punish you before death. But if you lived virtuously and worshipped properly, you would thrive.

As commentators have long noted, the Christian vision of heaven seems like a bore. You just sit around in the clouds listening to angelic harpists playing easy-listening for all eternity. It seems that it was hard for the early Christians, who tended to see natural appetites as shameful and regrettable, to describe a vividly incentivizing heaven. This is because, for most of us, true paradise is hard to imagine without at least a few deadly sins to indulge in (at a minimum lust, gluttony, and sloth). Nevertheless, living for eternity with the people you love, in perfect health and security, sounds plenty good compared to Hades.

But for all the bland satisfactions of heaven, it's the Christian stories of hell that are, and always have been, shiver-inducing in their raw carnality. Here's how Ehrman summarizes the torments of hell for ordinary sinners: "Blasphemers hanged by their tongues over eternal flames. Men who committed adultery are similarly suspended, but by their genitals. Women who have performed abortions on themselves are sunk in excrement up to their necks forever. Those who slandered Christ and doubted his righteousness have their eyes perennially burned out with red-hot irons. Those who worshipped idols are chased by demons off high canyons, time and again. Slaves who disobeyed their masters are forced to gnaw their tongues incessantly while being burned by fire."[14]

It's true that pagan mythology also describes the torments of the damned, but only for the most extreme sinners (e.g., Prometheus and Sisyphus are tortured for special defiance of the gods). But Christianity extended these extraordinary punishments to ordinary sinners: garden-variety liars, adulterers, robbers, blasphemers, onanists, and donkey-coveters.

And the whole Christian pitch was delivered with a tremendous sense of immediacy. Nothing about Christianity was remote. The pagan stories were awesomely durable, compelling enough to survive for countless generations as a kind of informal scripture and for millennia longer as great literature. But the pagan tales tended toward past-tense drama. *Christianity was a present-tense religion.* The miracles were occurring right now. And the early Christians insisted that they were living at the very climax of history. Jesus and his early followers expected the end of the world to come at any moment, likely in their own lifetimes. That the world would soon be coming to the end—with the virtuous rewarded and the sinners flung to hell—generated great urgency.

This was the "hard sell" element of the Christian pitch. Jesus could return today. Could return with your next breath. Unless you converted *right now,* it could be too late to save yourself and your loved ones from trillions of years of fiery torture.

Part of the secret to Christianity's early success, then, is stories of the afterlife that generated stronger activating emotions of surprise, awe, terror, dread, and hope. Research shows that these are exactly the type of strong emotions that erode skepticism, promote belief, and instigate the persuaded to go forth as persuaders. This helps explain how the Christian tales prevailed in the greatest (and still ongoing) story war of all time against spectacularly bad odds.

Speaking of narratives that thrive against apparently steep odds, let's shift now to conspiracy theories.

Crippled Minds

We're living through a pandemic of conspiratorial thinking.[15] Almost half of Americans believe in Area 51 conspiracies or are agnostic about them. Half of Americans believe in some version of 9/11 and JFK assassination conspiracies. A little less than a third believe in conspiracies about the New World Order and Obama birtherism. One-third of Republicans believe the QAnon theories about deep state elites are "mostly true." And speaking of pandemics, as I'm putting finishing touches on this book, support for Covid-19 conspiracies among Americans is surging: 40 percent think death rates have been "deliberately and greatly exaggerated" and 27 percent worry that Covid-19 vaccines might be used to plant tracking chips in our bodies.

The first and most important thing to understand about conspiracy theories is that the very term is a misnomer. The word *theory* suggests that, at bottom, these verifiably false narratives are believed because of a misfiring of reasoning faculties. But conspiracy theories, in all of their endless flavors, aren't about reason run amok; they're about powerful stories *causing* reason to run amok. So let's call these paranoid fantasies what they actually are: conspiracy stories.

The work of psychologists and other social scientists who study conspiracy stories is extremely valuable. But in their zest to break the psychology of conspiracism down to its building blocks, they generally overlook the simplest reason conspiracies catch on: because they're usually highly invigorating fictional thrillers. Almost all of the well-subscribed conspiracy stories could make blockbuster Hollywood films. Most conspiracy debunkings would only make okay PBS documentaries. Mainly, what debunkers have to offer isn't an equally gripping story but a story void. Debunkers face the killjoy task of showing that *there is no story*. There never was a summit of world Jewry where the elders kept minute notes on the world Jewish conspiracy. Area 51 isn't full of starships and dissected alien carcasses and scientists reverse-engineering alien laser cannons.

Conspiracy stories spread and thrive for the same reason disinformation of all kinds spreads approximately six times faster than truth on social media.[16] Conspiratorial fiction can be perfectly crafted to besot our imaginations, whereas true stories are always shackled by facts. Social media is a powerful if inadvertent experiment into what sort of narratives actually win out in story wars as measured by quantifiable views, likes, and shares. And this dynamic, whereby false information outcompetes the dull truth, mocks any faith that better information eventually triumphs in the marketplace of ideas and narratives. In truth, bad information—so long as it makes a good story—tends to outcompete a dull story packed with high-quality information.

Conspiracy tales are a prime example of Plato's warning that a rollicking good story can "cripple" our minds.[17] As we've already seen, highly transporting stories are also highly persuasive. "Researchers have repeatedly found," writes the psychologist Raymond Mar, "that reader attitudes shift to become more congruent with the ideas expressed in a narrative."[18] This all applies not just to novels and movies but to conspiracy tales that are usually more imaginative, juicy, and exciting than the truth.

But not always. The real moon landing story is far more amazing, heroic, and inspiring than the idea that the whole episode was clumsily fabricated on a Hollywood set. But as great as the true tale of the moon landing is, it doesn't *activate* consumers to move in the world. This is because the moon landing is literally history—it's over and done, and after reading Tom Wolfe's glorious book about the early days of the astronaut program, *The Right Stuff*, we say "how about that" and go on with our days. The true story of the moon landing demands nothing of the consumer but admiration.

But the moon landing conspiracy theory, in which NASA faked the landing for a variety of dastardly reasons, makes much stronger demands on the consumer. At the heart of this conspiracy theory, and all the other biggies, is an assertion of evil. Conspiracy stories are moral horror stories. And they're mostly written in the present tense. Although the moon landing or JFK conspiracies may have happened a long time ago, the dark forces that orchestrated them are still involved in mythical-scale evil. And all conspiracy tales call on their converts to do *something* about it as a matter of utmost moral obligation.

The first step in the hero's journey, folklorist Joseph Campbell's famous attempt to describe a cross-cultural structure of hero myths, is "the call to adventure."[19] There has been a disruption in the world, and the would-be hero is confronted with an urgent task—bad guys to fight, dragons to charm—in order to achieve some vital ethical end. Conspiracy stories operate the same way. By convincing formerly naive sheeple that hidden puppeteers are manipulating us all, conspiracy stories call would-be heroes to adventure. They call us to sleuth out clues that will expose the puppeteers. And they call us to brave the slanders of skeptics as we spread the bad news.

To get a better sense for why conspiracy tales are so infectious and resilient, let's consider one of the oldest, wildest, and most resilient conspiracies of all: that planet Earth is actually flat.

Pancake Earth

The flat earth movement is the brainchild of a roving lecturer, writer, and quack doctor named Samuel Birley Rowbotham (1816–1884), who went by the name "Parallax."[20] The Earth, Parallax declared, was not an awesomely ancient billiard ball, whizzing around the sun at sixty-seven thousand miles per hour with a wicked sidespin of a thousand miles per hour. The Earth was actually young, stationary, and pancake flat. The moon and sun moved in a continuous orbit over the pancake Earth like two little spotlights. The continents were clustered like berries at the center of the pancake. A huge ice wall encircled the perimeter like a squirt of whipped cream, which kept the oceans from spilling into the void. Parallax explained that no one was really sure what you'd find if you could actually cross the impassible wall of ice. But if you could slide a spatula under the pancake Earth, and give it a flip, you might well find hell on the reverse side.

Parallax wrapped up his flat Earth assertions in pseudoscientific terminology. But he stressed that you didn't need a fancy education to be a scientist. You didn't need to know math. And you didn't need any technical instrumentation. All you needed was your own common sense. Does it *feel* like our planet is whirling through space at thousands of miles per hour? Well, that's because it isn't.

Two percent of American adults, or roughly six million people, believe that the Earth is flat.[21] This might suggest that flat earthism is losing its story war in a rout. But from another point of view, flat earthism is punching far above its weight. The idea has lasted for more than 150 years despite its total scientific bankruptcy. And far more troubling, young people have been so inundated with well-produced and cannily argued YouTube videos, internet memes, and podcasts—not to mention a small flood of celebrity influencers—that a third of millennials say they're no longer entirely sure about the shape of the Earth. For an idea this dumb to capture any market share—while sowing doubt about the planet's shape in the minds of millions of young adults—is an *enormous* win for the movement.

How can a considerable number of people adopt a theory, flat earthism, that appears in dictionaries as a synonym for stupidity? The laziest explanation is that flat earthism is a shiny pseudointellectual bauble that draws in morons like a fishing lure. To the contrary, the leaders of the flat Earth movement strike me as intelligent people with lots of admirable intellectual qualities, including creativity, formidable powers of rhetoric, and truly impressive courage of opinion. In fact, research suggests that more-intelligent people aren't necessarily more rational, and they can be especially adept at spinning conspiratorial fantasies.[22] They use their superior intellectual horsepower to compose and elaborate the narrative equivalents of Rube Goldberg machines, which take enormous imagination to build and maintain. This all certainly applied to Parallax himself, who not only was smart but also had the rhetorical skills—and the duelist's temperament—to run circles around scientific authorities in debates.

The key thing to understand about flat earthers is they've never been drawn to the movement by scientific curiosity. A flat Earth society isn't made up of contrarian science nerds. When they fight for their views, they aren't really fighting over the shape of the Earth because . . . well, who cares? Flat earthers are partially right after all: the Earth isn't technically a ball, at least not a perfect one. Geologists describe the Earth as an oblate spheroid—smooshed slightly at the poles and chubby at the waist.

From the beginning, flat earthers have been story warriors, not science warriors. Parallax didn't *really* care about the shape of the Earth. He cared about promoting and defending his favorite story. And it wasn't just any story. It was the story of stories contained in the Holy Bible. Parallax was an adamant biblical creationist. We're used to modern creationists fighting evolutionary biologists over the origins and development of life. But Genesis tells the tale not only of the sudden origins of all life but also of the creations of the heavens and Earth. Parallax was a *geological* creationist. He felt that God created Earth in one day, and he interpreted a variety of scriptural passages as describing a flat Earth.

Parallax attributed the round Earth theory to scientific error, not fraud. But his followers soon charged that round earthism was a literally satanic conspiracy to hide the true shape of the world in order to shake faith in the Bible and promote a secular worldview.[23] Like modern biblical creationists, what flat Earth creationists were rejecting wasn't so much Earth science as the narrative flowing from that science. They were fighting the notion that the Bible was just a pile of dusty myths and that life emerged by chance, slithered out of the primordial slime, and evolved through a billion years of meaningless screwing and killing.

Over the last decade, flat earthism has enjoyed a great resurgence fueled mainly by digital media (YouTube's suggestion algorithm alone actively recommended flat Earth videos hundreds of millions of times).[24] Although some fundamentalist flat earthers continue to fight the satanic conspiracy, the most prominent flat earthers today are ordinary secular conspiracy theorists. And they face the same question of motive: Why would the Jews or the Bilderberg group or the Illuminati or Reptilian Overlords from the fourth dimension—to name a few of the likeliest suspects—work so hard to convince us that the world is round? It would take unbelievable discipline for the world's astronauts, scientists, sailors, airline pilots, mapmakers, and presidents—just to list some of the millions of people who would have to be keeping the secret—to pull the wool over our eyes. What's in it for them?

But on closer inspection, the secular flat earthers have a story that's big and compelling. The total effect is a lot like *The Matrix*, where you take the red pill and learn that everything you thought true is untrue and malign forces are controlling every aspect of your existence. Whereas fundamentalist flat earthers are living in a biblical epic, the secularists are living in a mashed-up sci-fi mystery thriller. Secularists are called to do the gumshoe work of piecing together the clues, exposing the perpetrators, and illuminating their hidden motives.

To attend a flat Earth convention is really not so different from attending a LARP convention. Only, rather than improvising their way through D&D-like scenarios, flat earthers get together to play the parts of heroic investigators. But flat Earth role-play is even more compelling because, unlike traditional live-action role-playing, which is hampered by a temporary and imperfect suspension of disbelief (everyone knows it's make-believe), the suspension of disbelief for flat Earth LARP seems to be authentic and complete. Inside the convention hotel, flat earthers aren't what society says they are: kooks, losers, virgins, ignoramuses. They're genius-level questers too far outside the box to be appreciated in their own time. But someday they will be recognized as heroes who answered the call to adventure and risked everything to expose the greatest fraud in history.

The ego gratifications of conspiracism help explain why, after a person takes up residence in conspiracy land, it's so hard to move them back to reality. As long as you stay within the conspiratorial narrative, you're a hero of thought. But to admit you're wrong is to admit you were inside a different story all along. The story wasn't a heroic epic featuring you and your friends charging down monsters. The story was a tragicomedy, and you were only charging windmills.

The Power (and Danger) of Quasi-Religions

Some psychologists categorize conspiracy stories as quasi-religions—parallel in form and function to traditional religion.[25] Although this comparison flatters neither the religiously minded nor the conspiracy minded, there's clearly something to it. The parallels between fundamental features of conspiracism and traditional religion are obvious and profound, suggesting that a culture's most and least august narratives emerge from the same facets of narrative psychology.

For example, both conspiracy stories like flat earthism and QAnon and a religion like Christianity emerged as word-of-mouth storytelling movements. Both enlist believers as protagonists in deeply meaningful crusades against evil. And both spread virally by inspiring activating emotions that drive adherents to spread the good (or bad) news with evangelical zeal.

Furthermore, both phenomena are characterized by a near-perfect invulnerability to disconfirming evidence. As is well understood of conspiracy stories, there's no evidence *against* the story that a believer can't creatively reinterpret as evidence *for* it. And this inoculation against disconfirming evidence is at least as impressive for religion. Up until just a few hundred years ago, most religious believers were fundamentalists who took their scriptures as literal truth. Then science patiently disproved most of scriptures' verifiable fact claims about the age of the universe, the formation of the planet, the workings of the solar system, the emergence of life, and so on.

Some believers remained fundamentalists, which meant rejecting this science as fake news. But most converted to figurative interpretation, saying, in effect, "Yes, the verifiable details in my infallible scripture have been proven untrue, but the religion is still 100 percent true. Who'd expect God to sink to pedantic literalism?" And when events occur that seem grossly at odds with the likelihood of a god who's both omnipotent *and* benevolent—such as when a couple of hundred thousand innocent people (mostly women and children) were killed in the Indian Ocean tsunami of 2004—the devout defend their faith with clichés ("God moves in mysterious ways") or they brandish the hammer in warning ("Who are *you* to question God's plan?").

Some religious people might absorb all this as an unprovoked attack. But it would be impossible to leave religion out of this book, given that it represents one of the purest expressions of the story paradox, with sacred narratives simultaneously driving great good and harm in the world. Moreover, my point isn't so much that believers in religious stories or conspiracy stories get overpowered by narrative. The point is that religions and conspiracy stories are just particularly vivid examples of general tendencies in narrative psychology. The narratives we use to make sense of the world are inherently greedy, in the sense that they always want to stretch out to explain more and more. And they are inherently arrogant, in the sense that they tend to deny their own flaws. And when a narrative bloats up with enough greed and arrogance, it becomes a master narrative—an attempt to explain virtually everything about the world. These master narratives, whether secular or sacred, are treated by their advocates as sacrosanct and unchallengeable, and when we do battle for these stories we do so as holy warriors on the side of right.

This is as true of dogmatic Marxists, ardent MAGA cultists, and the most primly pious of the "awokened" as it is of devout Christians and flat earthers.

Stories That Can't Win

This chapter has focused so far on stories that catch on against the odds. Now let's briefly shift to the opposite sort of stories: those that seem like they should catch on but don't. Why are some stories impossible for us to resist, no matter how unimportant, whereas other narratives can't break through even when our lives depend on it?

Toby Ord's book *The Precipice* (2020) encapsulates this problem. According to Ord's reading of the probabilities, exponential growth of human technology combined with stagnancy of human wisdom means it's increasingly likely that we will annihilate ourselves as a species, or at least crash civilization irrecoverably, by century's end. If sheer importance has any bearing on whether a book is read, this one should be a number one bestseller. But just a couple of months after its release, when most books are still selling near their peaks, this well-argued book was ranked 430,258th on Amazon's list of bestsellers.

Where do we place all the attention that we should be addressing to the issues covered in Ord's book, namely, climate change, bioweapons, killer plagues (well, SARS-CoV-2 has forced us to pay belated attention to this one), Earth-killing asteroids or volcanoes, the birth of potentially destructive forms of AI, and the fact that our entire planet is still rigged up with an insane nuclear doomsday system set on a hair trigger?

In 2007, while America was mired deeply in two foreign wars and hundreds of thousands were being killed, raped, or displaced by the genocide in Darfur, the world was meeting an ex–personal assistant and amateur sex-tape star named Kim Kardashian and breathlessly following her vapid semi-real soap opera on cable TV. Even people who've never watched an episode of *Keeping Up with the Kardashians* know a lot about the details of the characters' lives, whereas most Americans have hardly heard about Darfur, much less know where to find it on a map.

Coded deep into our DNA is a strong narrative bias. We attend to a narrative not on the basis of what's most important according to a rational or ethical calculation but on the same basis that we choose a movie for a Saturday night: What makes the best story?

I don't think this is because we want to hide from the world's misery and chaos inside escapist confections. Rather, it shows that our minds were not designed to deal with long-term, abstract threats like those covered in Ord's book. They were designed to deal with the daily eat-or-be-eaten—screw or get screwed—exigencies of hunter-gatherer life.

In the worlds of our ancestors, for example, the type of hot social information delivered by *Keeping Up with the Kardashians* would have been not only fascinating but also intensely important. Soap opera fodder about sex and conflict and shifting alliances bore directly on the tribe's ability to maintain harmony and resist fracture.

And while Ord's book is neglected, what's flying off the shelves? On the same day I checked Ord's Amazon rank, the number one bestseller in the world was a memoir/self-help guide by Glennon Doyle. Number two was the latest addition to Stephenie Meyer's teenage vampire saga Twilight.

This all raises the tragic possibility that because of the innate configurations of our storytelling psychology, we're unable to deal effectively with whole classes of problems that either (1) can't make a classically good story, or (2) make a good story but evoke the wrong type of emotion, the deactivating kind.

For example, a leading theory explaining humanity's sluggish response to global warming is that climate change happens to make a really bad story.[26] The best stories, the researchers say, hook us by dealing with clearly defined heroes and villains and dramatizing clear and present dangers—not a geophysical process that unfolds at a receding glacier's *drip-drip-drip* pace. Yes, you can tell a story about villainous oil executives and heroic environmental activists, but the main players are hard to characterize. They're either massive statistical clouds of human beings who are (like you and me) both the perpetrators *and* the potential victims, or they're the abstract forces of geophysics.

But the "bad story" theory of our feckless response to climate change overlooks the fact that people love tales of catastrophe and apocalypse. That's how Jesus himself moved his audiences—by telling stories of the end of the world. And end times mayhem is a favorite subject in blockbuster fiction about plagues of aliens, zombies, or microbes. We love stories of humanity up against the greatest threats conceivable. And not only does global warming fit this bill, the topic has featured in a number of popular films, from *Waterworld* to *Mad Max: Fury Road.* And there's now a whole genre of speculative fiction—cli-fi—dealing with the long-term challenges of climate change, with contributions by best-selling novelists like Margaret Atwood, Barbara Kingsolver, Kim Stanley Robinson, and Octavia Butler.

The problem with messaging climate change isn't that it makes an inherently bad story so much as an inherently deactivating one. Despite endless obfuscation by deniers, most people now accept the dire prophecies of climate science, and they can be engrossed by tales about it. But the scale of the problem is so vast and the obstacles thrown up by different governments, industries, and skeptics are so enormous that it's hard to see how we can ever come together as a human family to solve it.

In contrast to the abstractions of science, conspiracy stories about climate change can be highly activating because the good guys and bad guys are sharply drawn, and the problem is so much smaller. We aren't asked to solve the Gordian knot of scientific, political, and economic challenges associated with climate change. All we need do is stop the evildoers from spreading the hoax.

A Vast Conspiracy of Jolly Do-Gooders

The spread of conspiracy stories about global warming or anything else wouldn't be so bad if the stories weren't so *bad*. I don't mean they're bad in an aesthetic sense. I mean that the conspiracies that win out in the marketplace of storytelling are uniformly tales of bad people and bad tidings. There are no benign conspiracies. There are no cabals of shadowy philanthropists finding ingenious ways to enrich us in entirely unobjectionable ways. The chemtrails drifting overhead are never seeded with lifesaving medicines or herbal wellness enhancers.

The conspirators are never jolly do-gooders.

They're always monsters.

The reason conspiracies of goodness can't compete against conspiracies of darkness is simple. They make bad stories. They're too dull to seize attention and, because they don't pose a morally urgent problem, they don't call on us to enter the tales as heroes or, at least, to open our mouths and breathe them around like viruses.

All of this connects to a general pattern in the story wars whereby darker material usually beats out brighter material. And this pattern is, in turn, driven by a broad "negativity bias" in human psychology, by which, as the psychologist Daniel Fessler and his colleagues have shown, "compared to positive events, negative events more readily capture attention, are stored more readily in memory . . . and have greater motivational impetus."[27]

This negativity bias is reflected in humanity's innate and highly restricted palate of narrative tastes. Storytellers must adhere to this palate in the same way that chefs must conform to the five basic tastes when cooking our food. The upcoming chapter expounds on these narrative regularities that stretch back thousands of years to the first recorded stories and almost certainly to the oral tales of prehistory. These regularities drive the joy and goodness of story-telling but also unleash whirlwinds.

4

THE UNIVERSAL GRAMMAR

IMAGINE THE NOVELIST JAMES JOYCE PEERING THROUGH HIS WINdow into the Parisian night.[1] He's chain-smoking. He's sniffing medicinal cocaine for his ailments. He's adjusting his Coke-bottle glasses and squinting down at his big notebook pages as he scratches out *Finnegans Wake* in blue crayon. He's laughing so hard at his own jokes—his rollicking wordplay, his lewd asides—that his long-suffering wife, Nora, shouts at him from bed to knock it off so she can get some sleep.[2]

The writing of *Finnegans Wake* was an act of breathtaking literary swagger. Like Jackson Pollack slinging paint at canvas, Joyce wanted to smash the traditional grammar of his art form. Frustrated with the limitations of English, he invented his own language, mashing together words and word bits from different tongues into a new dialect. Bored of the contrivances of "cutanddry grammar and goahead plot," Joyce mostly did away with plot.[3] And while he was at it, he obliterated the notion of character, too. Joyce's characters shift and morph, changing names, personality attributes, and physical traits. James Joyce set out to take something as old as humanity—the storytelling impulse—and make it new.

I'm in awe of this book even if, in the half-dozen attempts I made as a younger and braver man, I've never been able to read that much of it. To see why, just try to read the first four sentences of *Finnegans Wake* below, and then imagine summoning enough masochism to grind through seven hundred pages of the same:

> riverrun, past Eve and Adam's, from swerve of shore to bend of bay, brings us by a commodius vicus of recirculation back to Howth Castle and Environs.
>
> Sir Tristram, violer d'amores, fr'over the short sea, had passencore rearrived from North Armorica on this side the scraggy isthmus of Europe Minor to wielderfight his penisolate war: nor had topsawyer's rocks by the stream Oconee exaggerated themselse to Laurens County's gorgios while they went doublin their mumper all the time: nor avoice from afire bellowsed mishe mite totauftauf thuartpeatrick: not yet, though venissoon after, had a kidscad buttended a bland old isaac: not yet, though all's fair in vanessy, were sosie sesthers wroth with twone nathandjoe. Rot a peck of pa's malt had Jhem or Shen brewed by arclight and rory end to the regginbrow was to be seen ringsome on the aquaface.
>
> The fall (bababadalgharaghtakamminarronnkonnbronntonnerronntuonnthunntrovarrhounawnskawntoohoo-hoordenenthurnuk!) of a once wallstrait oldparr is retaled early in bed and later on life down through all christian minstrelsy.

Joyce—mostly blind, toothless, obsessed with his money and his fame—worked heroically at *Finnegans Wake* for seventeen years and bragged that it would keep literary critics busy for three hundred more.[4] In this, he probably succeeded. The book is now hailed as a towering monument of experimental art and as one of the greatest novels ever written. According to Yale critic Harold Bloom, *Finnegans Wake* is the one work of modern literature whose genius stands comparison to the masterpieces of Dante and Shakespeare.[5]

But there's a paradox about *Finnegans Wake*: It's known as one of the greatest novels a human has ever penned. But it's also a novel that's so maddeningly hard and weird (one critic refers to it as an act of "linguistic sodomy")[6] that almost no one can stand to actually read it. I'm a literature PhD and, aside from a Joyce scholar I once knew, I've never met a single colleague who claims to have read the whole thing or has been greatly tempted to try.

Finnegans Wake's failure to attract actual readers, which seemed to genuinely surprise and disappoint Joyce,[7] reflects a foundational truth about the art of storytelling: it works inside fairly narrow bounds of possibility. It's not something that can be endlessly rejiggered and reengineered.

We can imagine narrative transportation as an especially vulnerable brain state that's protected by a lock that tumbles open only to specific combinations. For as long as there have been humans, the ways of undoing the lock have been passed down through generations of storytellers. Going back to the earliest forms of oral folktales and moving forward through stage plays, printed histories, and modern YouTube shorts, *the fundamentals of successful storytelling have not changed at all.*

The linguist Noam Chomsky theorizes that all of the thousands of languages spoken on Earth share a universal grammar. The tremendous surface variations among languages disguise commonalities that emerge from the universal design of human brains.[8]

Chomsky's universal grammar has always been controversial, with experts split for and (now increasingly) against. But the idea of a natural grammar of storytelling should be much less controversial. The grammar of language is wickedly hard. When encountering a foreign tongue, we must take extreme pains to gain minimal comprehension, to say nothing of mastery. But the grammar of stories is easy by contrast. Once stories are translated, and once any unfamiliar cultural matters are explained, stories from foreign lands are almost always effortlessly comprehensible and enjoyable.[9]

It's this natural grammar of story that spins the tumblers of our mental locks and gives us the joy of story. When storytellers defy this universal grammar, they're still making noise, and sometimes still making sense, but they're usually making avant-garde word art, à la *Finnegans Wake*, not stories. Avant-garde word art is fine if that's what you're into. But not if you want to attract an eager audience and draw them into the trance of narrative transportation.

The universal grammar of storytelling has, I propose, at least two major components. First, everywhere in the world stories are about characters trying to resolve predicaments. Stories are about trouble. Stories are rarely about people having good days. Even comedies, though they often end happily, are usually about people gutting through bad days—often the very worst days of their whole lives. Second, as corny as it may at first sound, stories tend to have a deep moral dimension. Although sophisticated novelists, historians, or filmmakers may deny that they'd ever sink to expressing anything like a "moral of the story," they've never stopped moralizing for a moment. "The poets," says Nietzsche, "were always the valets of some morality."[10]

Stipulated: If you ransack your brain, you'll be able to name exceptions. But they will be exceptions that prove the rule—statistical outliers dominated by experimental works like *Finnegans Wake*. On the other hand, maybe it seems obvious to you that stories *are* this way. But academic literary theorists would mostly deny it.[11] And, if you think about it, it's not a bit obvious that stories *should* be this way. Many of us might expect to find storytelling traditions where stories mostly function as escape pods into hedonistic paradises where pleasure is infinite and moral trespass is unknown.

We never do.

The Anguish of a Happy Ending

Never? Well, it depends on how strictly you define the word *story*. Consider porn, which is among the most dominant forms of storytelling in the world.[12] Porn in the film and VCR eras mainly unfolded at feature length and slavishly aped (or spoofed) cinematic conventions and story scenarios. Now, most porn is consumed in quick and dirty clips. But even these clips usually have a minimal narrative basis. The clip first establishes a fantasy scenario: your step-sibling is so attractive and it isn't *really* incest after all; frustrated wifey aches for a *real* man. Porn stories may be rudimentary, but they aren't actually superfluous.

But porn is nonetheless the one form of popular narrative (or narrative-ish) entertainment that places the consumer in imaginative scenarios of pure wish fulfillment. Other forms of storytelling, with their focus on conflict and struggle, are different from porn except in one overlooked sense. Once the climax is reached, porn and mainstream storytelling both quickly cut to the credits. In porn stories, as in every other type, once the tension is released, no one wants to hang out basking in the afterglow.

The little truism *everyone loves a happy ending* reveals a big, confusing paradox. Taken literally, it suggests that people like happiness in their stories, but only at the end. The 2018 blockbuster film *A Quiet Place* flings viewers into a hellish nightmare where the world has been depopulated by an invasion of otherworldly carnivorous bugs. The bugs are hunting a young family, devouring them one by one, until two of the surviving family members, the mother and a daughter, suddenly discover the monsters' fatal weakness. The film's happy ending—its entire shift from nightmarish torture simulation to triumphant resolution—lasts precisely as long as it takes for the women to exchange a single glance. Their eyes flash with all their hope of life and all their lust for revenge.

It's a stylish, perfect happy ending in that it gives viewers all the good feelings we like at story's end while sparing us the tedium of lingering in the newly hopeful world for a split second longer than necessary.

A Quiet Place illustrates the queer eagerness humans have to actually pay for narrative simulations of horrors that we'd pay anything to avoid in real life. But it's also an illustration of an even more dependable pattern in stories. Long novels, films, and TV series resolve the central problem in the last few pages or minutes, and then typically end very quickly. For example, the main formula for episodic TV is that each chapter is about problems and conflict until the very end, when the action moves quickly through resolution to the credit sequence. Whether the struggles and victories are small and silly, as in a sitcom, or deadly serious, as in a crime procedural, we know the next episode will find our heroes embroiled, yet again, in struggles climaxing in literally last-minute resolutions.

After subjecting summaries of 112,000 fiction plots to statistical analysis, the data scientist David Robinson reached the following pithy conclusion: "If we had to summarize the *average* story that humans tell, it would go something like, **Things get worse and worse until at the last minute they get better**" (his emphasis).[13] Even news programs hew as closely as they can to the formula. They focus overwhelmingly on stories of real-world problems and then give us one "fluff" story at the end—sending us away with a burst of hope.

This isn't a modern convention. After all, the very phrase "happily ever after" is taken from traditional folk and fairy tales. "Happily ever after" is how folk storytellers fast-forwarded through the dullness of a blissful resolution, which allowed them to move on to the next tale of cannibalistic witches, rampaging giants, and evil stepmothers.

Why? Why is there so little interest in happiness? Why does it bore us so? Once the bad guys are defeated, why don't we yearn to tarry in those storylands of streets made safe and true love and crops of happy children?

Whatever the answer, it suggests that narrative psychology has a sadomasochistic bent.[14] Storytellers have to linger on what's bad to make us feel good. And as soon as things turn nice in our story worlds, we want out fast. And where do we want to go? Sometimes we want to return to reality. But as often we turn the page to the next story—or click on the next program in our queue—and dive back into a new world of struggle.

It Isn't Good Unless It's Bad

Going back at least to the works of Shakespeare, we've understood the social purpose of the storyteller as holding "the mirror up to nature."[15] A proper story reflects the truth of ourselves and our times with clarity and fidelity. But what if the mirror of storytelling is actually badly warped? And what if, even worse, the mirror is warped in exactly the same way irrespective of the teller's politics, gender, ethnicity, historical circumstances, and preferred genre? What if the mirror perversely fuzzes out most of what's pretty and harmonious in the world, while accentuating blemishes and sores? And what if, finally, thanks to the innate gullibility of narrative psychology, we're strongly predisposed to believe in the distorted image?

Take the most extreme sort of negativity conceivable: death. A recent study by the psychologists Olivier Morin and Oleg Sobchuk concludes that, if you're unlucky enough to be a character in a diverse sample of 744 twentieth-century American novels, your odds of dying—often suddenly and hideously—are higher by at least an order of magnitude than a real person's.[16] But story's bias toward problems and turmoil extends well beyond the boundaries of fiction to characterize every genre of tale-telling, from casual gossip to conspiracy stories, from journalism to political campaign narrative, from the threadbare anecdotes traded around at parties to the make-believe of little children.

In historical storytelling, for example, as in every other kind, accounts of conflict, chaos, pain, oppression, and death always exert far more suction on authors and readers than narratives of peace and harmony. This is true for the fussy scholarly histories published by academic presses, but it's far more extreme for the histories ordinary people consume in books and documentaries. If you doubt this, go to your local bookstore and browse the shelves. You'll find that the subject matter on the history shelves is every bit as chaotic and conflict-ridden as the stuff on the fiction shelves. In fact, writers are unlikely to prevail in the fierce competition for space on those shelves *unless* they have the good business sense to write about episodes of conflict and struggle.

This makes history, taken as a whole, a picture not of life as it really was but of life passed through a screen that filters out most of the good times in favor of war, famine, death, plague, and other stuff that makes good drama. The negative focus of historians is captured in George Santayana's famous slogan: "Those who can't remember the past are doomed to repeat it."[17]

What's the past, according to Santayana? It's a story of bad men and bad tidings, which would be doom to repeat. But that's not really how the past was. It wasn't all pogroms and witch burnings, nor was it as nasty, brutish, and short as our histories might lead us to believe. History's negative fixation clouds our ability to see the past as it really was not just in the worst times but in the best, and therefore to learn lessons that might help us calibrate our responses to present problems and opportunities.

The mirror that journalists hold up to the world is warped in precisely the same way. Walter Cronkite used to sign off his nightly news broadcast with "And that's the way it is." But, no, it wasn't! Cronkite claimed to give us a clear picture of the world, but he actually presented a kind of reality show where the most troubling news of the day was edited down into a comprehensive misrepresentation of reality.

What the historian Robert Muchembled has said of the origins of the newspaper industry in fifteenth-century Europe is as true of the news industry today: "Blood and gore sold ink and paper."[18] From the earliest news pamphlets printed on the earliest presses, news has always suggested that the world is crap and getting ever crappier.[19] But in two books of great heft and importance, Steven Pinker deploys loads of data in support of an idea that many people find literally unbelievable: the world is better than it has ever been, and getting still better all the time, in almost every way that can be measured.[20] Violence is greatly diminished, both in terms of interpersonal conflict and the destructiveness of war. The burden of racism is massively reduced, and sexism, too. Bullying is way down, as is the abuse of children and animals. Pollution is down. Human rights are up. Medical care is better than ever. Even poor people are so much better off than yesterday, and so much better fed. Almost everything is getting better and few things are getting worse. (Though some things are, like political polarization.[21] More on that later.)

This doesn't mean we live in utopia or that we shouldn't worry about our problems. Moreover, it doesn't mean we're anything like safe: an asteroid might destroy us, or we might slowly cook ourselves with our emissions or flash-fry ourselves with nukes or just drive ourselves fully mad with stories. In fact, as technology grows ever more powerful and democratized, we face a level of existential risk we never faced in the past.[22] But Pinker's data do show that our simple "hell-in-a-handbasket" intuitions about the world are wrong. And most people can't see this because journalism passes reality through a filter that captures and discards good news while amplifying bad.

I'm not the first, or millionth, person to notice this. The media scholar Betsi Grabe finds that "compelling negativity is the most persistent news selection principle—over time, across cultures, and despite damning criticism."[23] Why, as Pinker puts it, does the news depict the world as "a vale of tears, a tale of woe, a slough of despond" even as the objective facts about the world have brightened considerably?[24]

Journalists feel a bit guilty about this, but they have a stock answer at hand. It's not news's job, they say, to tell happily-ever-after stories about the world. That's make-believe's job. The job of the news is to sniff out problems. After all, we can't fix the world if we don't know what's wrong with it.

This is a good story in the sense that it has a certain ring of plausibility. But the heroic story journalism tells about itself is, as they say, only *based* on a true story. Journalism is a storytelling guild, which, like all other storytelling guilds, is trapped helplessly inside the universal grammar of storytelling, with no plausible hope of escape. The history of the news business shows that there is, in fact, little market for news per se, and never has been. There's only a market for drama. And, from the beginning, the news business has been just one branch of the drama business, where the elements of the universal grammar—the plots, characters, themes, and implicit ethical lessons—must be cobbled together from real life.

Good drama, not representative truth, is the main criterion of "newsworthiness." And good news makes bad drama. Although we like happy endings, a story generally isn't "good" if it's smooth sailing all the way through. A journalist may do fine reporting and fine writing on a story that's important in an objective sense, but if the story doesn't highlight conflict, we won't find it to be good. When it comes to stories, almost nothing's good unless it's bad.

As Plato himself said, "Virtue is not necessarily the best choice of subject for a man who wants to write a beautiful epic or drama; the poet must subordinate his love of virtue to the requirements of his art."[25] And the same goes just as strongly for a news story as it does for fiction. For individual journalists, telling happy stories isn't a high-percentage strategy for keeping a job much less winning a Pulitzer. And for entire news organizations it would be bad business to prioritize good news for exactly the same reason it would be bad business for HBO or Netflix to dial down the conflict and turmoil.

The extreme negativity of the news has serious consequences for the world. In fiction things typically get worse and worse until, at the very end, they get better. Because made-up stories tend to end happily, psychologists find that heavy fiction consumers, when compared to heavy news consumers, have greater confidence that they live in a "nice world" rather than a "mean world."[26] They're more likely to think that the world is a good place where things turn out well in the end. Perhaps this means that fiction turns people into suckers. But just as plausibly it turns them into nicer people who believe that good people can overcome daunting obstacles to improve the world.

News stories, on the contrary, typically start mean and end mean, too. Journalists can't engineer happy endings—they can't deliver the comfort that, no matter how bad things may look, the good guys will triumph in the end. If the happy endings of fiction mostly produce optimism, news stories produce pessimism, paranoia, hopelessness, and emotional deactivation.

The point here in not to attack the news. Journalism performs an absolutely essential, and frequently heroic, social function as a de facto fourth estate. But like every other sort of storytelling, journalism has great capacity to do harm as well as good. The news is supposed to calibrate us for rational behavior. But it miscalibrates us in crucial ways. News consumers get an overall message that the world is an unsolvable mess—lock down your kids, strap on your guns, flee from the city to the suburbs, and from the suburbs to the survivalist compounds.[27] And while we are at it, let's elect a strong man to keep us safe.

Story's universal grammar conditions us to see the world in terms of proliferating problems and then to name, shame, and punish the villains who are responsible. The consequences of negativity bias can be bad, but the consequences of this moralistic bias are worse.

The Everlasting War of the Princesses and the Tigers

The two little girls had tired of their princess play, but as one swung on the swing set and the other pushed, they were still wearing their frilly princess gowns and sparkly tiaras.

The blonde one hopped off the swing and landed barefoot in the grass. "Let's pretend," she said, turning to her friend, "that our parents are dead—bited by mean tigers."

And just that fast, the bland suburban backyard with its fence and soft grass was transformed into a forest teeming with threats. The girls squatted low. They squinted along the fence line and saw predatory eyes gleaming through the bushes. They saw the tiger stripes rippling. And then the two friends ran.

I watched them from the back porch, fascinated. The blonde one was my younger daughter, Annabel, six years old at the time. She could have spent that afternoon in an imaginary paradise of unicorns and curly fries and fairy dust. Instead, in unconscious obedience to the universal grammar, she dreamed herself into the most intense horror scenario her six-year-old mind could conceive. Tigers had devoured the two girls' loving parents, and now they were hopelessly lost in the forest, with no adults to protect them and with the still-famished tigers hot on their trail. In this fantasy, all of life shrank down to fight or flight. The princesses struggled. They thought up clever ways to hide, clever ways to find food. And if they faltered, even for a moment, the tigers would gobble them up.

But what captured my attention wasn't just the horror of Annabel's imaginative scenario but its inchoate moral structure. She called her make-believe scenario "Lost Forest Children." But in the notes I jotted afterward, I called it "The Everlasting War of the Princesses and the Tigers" because it drove home the fact that most tales are battles between princesses and tigers—between the good guys and the bad. And as in Annabel's playtime scenario, which she spun up again and again that summer with different friends, the war never ends. The war between the good guys and the bad must be fought out in humanity's stories again and again until the end of time.

Annabel's story wasn't merely about survival. Primal if unspoken values animated the tale: the love of families, the need for people to cooperate, and the human necessity of finding refuge in community. For any of this to happen, the greed and savagery of the tigers had to be foiled. It was hard for my daughter, young as she was, to imagine that tigers might eat her parents with as much malice as she might munch a grape. In her imagination, the tigers were archetypal bad guys. They were monsters who preyed on people not out of amoral, biological necessity but because they were "mean."

And at the core of the drama was the bond between the little girls who faced monsters together and survived against impossible odds, all because they were good guys who never turned their backs on each other and always called out warnings and packed leaves into each other's claw wounds and never ran ahead to save themselves when the other tripped on the hem of her gown but who always risked everything to save each other, sometimes driving the tigers back by swinging a fallen stick like a sword or hurling it like a spear.

And as I watched, I was moved by the beauty of this ancient and instinctive simulation of humanity's greatest hope and deepest fear: by loving each other and sacrificing for each other, we might *live* together instead of perishing alone.

But there was a dark potential running through this story, too—through most stories. Love is generated, but so is hate. Community is forged, but enemies are identified. Empathy is created, but also callousness for those who haven't earned it. This suggests that stories are themselves like princesses, weaving us together into stronger collectives. But they can also be like tigers that claw out the stitches and leave us in tatters.

The volatility of stories, their often-chaotic effects on social life, comes down to three connected truths:

People need stories.

Stories need problems.

Problems need villains to cause them.*

Deus ex Machina

Stories, like real life, are full of luck—wild coincidences and unlikely tumblings of probabilities. But here's the difference between how luck typically circulates in stories versus real life. In real life, luck doesn't just precipitate events; it determines outcomes. Very good or very bad luck splashes into our life stories at the beginning, middle, and end. But a story, to be satisfying, generally has to control the role that luck plays in outcomes. As a general rule, luck plays an outsized role in the setups of stories while playing a much smaller role in resolutions.

Harry Potter isn't going to defeat Voldemort, for example, because the latter slips on a banana peel and breaks his head. True, Harry's friends often rush in to save him in the nick of time. But although this timing may be fortuitous, the fact that Harry and his friends risk their lives for each other says far less about random outcomes and far more about the values of loyalty and self-sacrifice that mark the good guys of storyland.

Deus ex machina is Latin for "god from the machine." In ancient drama, characters dressed as gods would sometimes be lowered onto the stage by a "machine," typically some sort of crane. Then

* Well, *usually* they do. That stories sometimes get by without villains of any kind is a hopeful point I'll take up later in the book.

the god would march around the stage setting things right—casting down the wicked, exalting the virtuous. This ancient term has since been expanded to encompass all kinds of lazy, ham-handed storytelling where happy endings are engineered through highly improbable occurrences.

In fact, Harry triumphing thanks to a banana peel would be *more* resented than the conventional deus ex machina descending from the sky. At least a deus ex machina would send a message about an intensely meaningful universe, ruled over by wise and benevolent deities. On the relatively rare occasions when luck plays a truly decisive role at the end of a tale, the teller is either inept or is trying to convey a very particular type of meaning: there is none.

This carefully controlled role that luck plays in story characterizes most fiction. As the novelist and creative writing teacher Steven James puts it in advice to aspiring writers, "Coincidence is necessary to get a story started but is often deadly at the end. However, too many authors use it backward: They work hard to get readers to buy into the plausibility of the beginning, but then bring in chance or convenience at the climax—when readers' coincidence tolerance is at its lowest."[28]

But this all applies to satisfying stories told in any genre. To take just one example, in *Thinking Fast and Slow* (2011), the psychologist Daniel Kahneman describes how the role of luck is utterly discounted as a determinative factor in storytelling about business: "Stories of how businesses rise and fall strike a chord with readers by offering what the human mind needs: a simple message of triumph and failure that identifies clear causes and ignores the determinative power of luck."[29]

A story is a way of structuring information—whether factual or invented or something in between—so as to produce and convey meaning. But what sorts of meaning are stories trying to deliver? Of course, the answers would seem to be as wildly various as the personalities and obsessions of the world's storytellers. But look closer and you find that stories from *Macbeth* to *MacGyver* are obsessed, above all, with the problems of right and wrong, good and bad.

The great science fiction writer Kurt Vonnegut argues that stories have a few standard shapes, one of which is called "man in a hole."[30] But there's a sense in which almost all stories are about people who fall into holes of one kind or another and have to struggle their way out. And I'd add that protagonists aren't often alone in the hole. They're sharing it with representations of human darkness. This could be the garden-variety smugness or selfishness confronted in a novel of manners or the capital *D* Darkness of *Star Wars* or the absurd, lowercase villainy confronted in a Will Ferrell comedy. Moreover, the darkness may be represented by a villain or it may be an inner darkness that the protagonist struggles to defeat (as in an antihero story).[31]

If there's anything that stories are *mostly* about, it isn't violence, sex, survival, power, or love—it's the abstraction of justice. People have insatiable appetites for structurally monotonous tales of moral trespass and the pursuit of justice. The nail-biting emotional tension stories build and then resolve is most frequently driven by our hopes for poetic justice: Will the characters get what they deserve? Will the good guys be rewarded? Will the bad guys be thwarted?

The centrality of poetic justice isn't an idiosyncratic convention of Western culture. As the literary scholar William Flesch argues in his book *Comeuppance*, it's a cross-culturally and cross-historically predictable feature of tale-telling. Even tragic stories, as the anthropologist Manvir Singh writes, are often given stealthy happy endings by offsetting "the protagonist's misfortune with redemption. *Romeo and Juliet*, for instance, ended not with the star-crossed lovebirds' deaths but with their families concluding their feud."[32] In a similar vein, James Cameron's *Titanic* seems like it is going to end in the most tragic way: the hero dead, the heroine crushed by trauma, and the villain bullying his way to safety on a lifeboat. But we learn that

Rose goes on to have a life rich with adventure and love, the villain gets an awful "ever after" ending in his suicide, and the film concludes with a vision of the protagonists reunited in a heavenly vision.

Of course, like all theories or generalizations about a topic as vast and baggy as storytelling, there's a need for caveats. First, I'm not speaking in absolutes. I don't claim that you'll never find examples of stories that break out of these patterns. After all, many artists strive for originality—they try to identify templates so they can smash them. I'm saying that stories that violate these patterns are ripples running against a heavy tide. And, at the risk of stating the obvious, when storytellers do opt to swim against the tide of poetic justice, they often provoke the ire of their readers and viewers. "In a number of experiments," writes the psychologist Dolf Zillmann, "it has been found that people experience pleasure when a liked character behaves well and succeeds. People experience frustration and anxiety when a disliked character behaves badly and succeeds."[33]

Furthermore, I'm claiming that this moral structure gives audiences such deep emotional satisfaction that it will rarely be absent from the most popular—and therefore the most influential—tales. This is supported by studies finding that the higher the level of poetic justice in a TV show the better it performs in the Nielsen ratings.[34] And it's also supported by the grandest and most influential tales people tell, our religious myths. The world's most successful faiths—Christianity, Islam, Hinduism, and Buddhism, which together have cornered 76 percent of the world's faith market—embed their adherents in mental simulations of perfect poetic justice. In these storylands no one gets away with *anything*. No vice goes unpunished, no virtue goes unrewarded, whether in this life or the next.

Graphing Jane Austen

When I express these ideas, people reliably resist. This is especially true of film and literature aficionados, who desperately want to believe that the craft of story is too subtle or rebellious to be constrained by laws of aesthetics. So let me tell you how I, along with some collaborators, began to develop what may at first seem like an overly simple way of thinking.

In the mid-2000s I set out, along with my colleagues Joseph Carroll, John Johnson, and Dan Kruger, on a large-scale study of classic Victorian novels by such authors as Jane Austen, George Eliot, and Charles Dickens, among many others. We distributed a survey to hundreds of knowledgeable people—professors, graduate students taking courses on Victorian literature, and authors who had published articles or books in the field. The respondents rated the attributes of characters in the novels exactly as if these fictional people were actual people.

We wrote up the results in our book, *Graphing Jane Austen: The Evolutionary Basis of Literary Meaning*. The main finding had to do with something we called agonistic structure, which we took to be a fundamental structural element of storytelling in much the same way that roofs are fundamental to houses. For all the variety in these novels, for all the differences in personality and gender and background of authors stretching over a century, they made strikingly similar choices regarding characterization. As a whole, Victorian novels reflect a sharply polarized fictional universe of good people (the protagonists and their allies) and bad people (the antagonists and their allies) locked in conflict. Overwhelmingly, protagonists looked to cooperate and work toward the common good while antagonists sought to dominate for selfish ends.

When we published *Graphing Jane Austen*, many readers were thunderstruck. Only not by the results themselves—but by the fact that a team of PhD scientists and scholars would dare to publish such obvious findings. After all, Victorian novels have a reputation for priggishness. But we were thrilled with the results because they actually do portray something anomalous. We're all just so desensitized to the anomaly that it hides in plain sight.

From one point of view, it's obvious that, despite exceptions, most stories portray "goody-baddy" dynamics—from nursery rhymes to juicy gossip, from ancient folktales to Holy Scripture, from lowbrow reality shows to award-winning documentaries.

The question is, why?

The *Graphing Jane Austen* team was inspired by the work of the anthropologist Christopher Boehm, who gained renown in the early 2000s for demonstrating that the lives of hunter-gatherers were typified not by the every-man-for-himself behavior that most associate with Darwinism but by a much gentler ethic of communalism and egalitarianism.[35]

The golden rule of hunter-gather life is pretty simple: Do everything to bring the band together; do nothing to split it apart. Don't sow division. Don't hog up more than your share (of food, sex partners, attention). If you happen to be blessed with muscle, don't throw it around. If you happen to be a great hunter or a dazzling beauty, don't flaunt it over others. Be one of the good guys, in other words.

Of course, if there's an instinct in humans to get along with each other in groups, there's also an instinct to get ahead. We proposed that agonistic structure in stories generally, not just in Victorian novels, reflects the ancient morality of hunter-gatherer life. Living in groups, as humans must, means constantly balancing our selfish impulses with group needs. Protagonists of stories properly balance individual self-interest with the needs of the group. In general, protagonists sacrifice their self-interest for the common good when the two are in conflict. Not antagonists. Pretty much the definition of a bad person among hunter-gatherers is the bad team player who reliably puts his or her own interests above the group's.

Drawing partly on our work in *Graphing Jane Austen*, the communications scholar Jens Kjeldgaard-Christiansen builds an instruction manual for creating powerful antagonists: "Evolutionary psychology supplies a basic blueprint for impactful villains: they are selfish, exploitative, and sadistic. They contravene the prosocial ethos of society." Giving a little creative writing advice, he continues, "Antagonists should be hyper-individualistic bullies. They should threaten the social order and induce righteous indignation in protagonists, incentivizing them and their peers to band together, fight back, and finally affirm their prosocial values."[36]

Kjeldgaard-Christiansen is describing the archetypal villain, the one who has been haunting stories in different guises since the origins of humanity. The villain is always the same deep down: he's the guy who wants to split us apart. She's the selfish one at work. The ball hog at the pickup game. The jerk who steals your parking spot. The murderer hiding in the bushes. The selfish bastard who's going to subvert the egalitarian ethos that holds the community together.

And who's the hero? The term *hero* evokes ancient associations with physical muscle and physical courage. But with much greater predictability, protagonists embody moral virtues, not physical ones. The protagonist is rarely a saint and not just because saints are boring. An interesting protagonist must have room for improvement. Creative writing teachers sometimes call the main protagonist of a story "the transformational character." Main antagonists usually don't evolve. Main protagonists do. In most cases, the transformation is moral. Protagonists go from being takers to givers. From blindness to sight. From confusion to understanding.

Storytelling is then—in every era and every culture—a dramatization of the everlasting war between the princesses and the tigers. Protagonists struggle to lay down the stitching that binds families, friends, and communities together. Antagonists seek to find those seams and rip them apart. Stories unfold as races: Who can do their work better and faster, the princesses or the tigers? The weavers or the cleavers?

In most tales, the protagonists struggle hard and eventually win out. But the larger war between the human animal's best and worst angels is ultimately unwinnable. And so the archetypal heroes and villains must be resurrected again and again to do battle until the end of time.

Stories Make Tribes

In *Graphing Jane Austen*, we argued that storytelling—like other forms of human art-making—has deep roots in the well-being of tribes as well as individuals. Story is one key solution to the problem of maintaining cooperation and cohesion within human communities. The moralism we see in our tales doesn't just reflect our evolved morality—it strongly reinforces it.

My colleagues and I aren't alone in these views.[37] In his book *Grooming, Gossip, and the Evolution of Language*, the primatologist Robin Dunbar argues that human language first evolved for the purpose of telling stories—namely, gossip tales about who was playing by tribal rules and who wasn't. Dunbar and other scientists stress that gossip, although it has a bad reputation, helps a community function by policing moral infractions. If Dunbar is right, this goes a long way toward explaining humanity's ongoing and probably incurable addiction to moralistic stories that are so like the gossip tales we've been sharing since our species first learned to talk.

Dunbar has since studied story's capacity to weave groups together with biochemical thread. He has shown, for example, that when we watch an emotionally arousing dramatic film our nervous system releases the endogenous opioids known as endorphins—this is the same neurobiological mechanism known to underpin bonding in humans and other primates. After viewing an emotionally intense drama, viewers have higher endorphin levels and report feeling a stronger sense of connection and belonging with the people around them.[38]

Dunbar proposes that storytelling is just like other art forms, such as singing and dancing, that pull us together in groups, trigger a similar release of bonding chemicals in the brain, and enhance a sense of group solidarity. Nowadays, we mainly consume stories alone or with our immediate families. But recall that up until the invention of the printing press around six hundred years ago most formal storytelling was oral and often festive and it was performed for groups of people by tellers or actors. In other words, story was an overwhelmingly communal activity, where people were transported together into storyland to experience the same simulations of morally charged scenarios and the same floods of ideas and emotions.

While Dunbar studies the neurobiology of stories in the lab, the anthropologist Daniel Smith and his colleagues have begun studying its effects in the field. In the study of Agta hunter-gatherers mentioned in Chapter 1, researchers tested the idea that story evolved to promote cooperative norms in groups. They found not only that the Agta put a premium on good storytelling but also that the prosocial messages in the tales actually sink in and modify group behavior. Agta bands that boasted strong storytellers also functioned better as cohesive and harmonious teams.[39]

In the spirit of a picture being worth a thousand words, this whole hypothesis is neatly encapsulated in the image of the Khoisan storyteller from Chapter 1. Flip back to the image one more time to marvel at the tribal unity the teller has generated. He's brought his people together, skin up against skin, mind up against mind, as he spins a tale pregnant with lessons that will promote the band's flourishing. I don't linger on this image because it represents "primitive" people involved in a primitive behavior (the Khoisan aren't primitive). I linger on this image because it's an utterly timeless expression of story's ongoing role in weaving individual people into functional collectives.

Not Moral, Moralistic

Tellers and critics who reflexively bristle at the very mention of a "moral of the story" are, I think, actually objecting to two different things. First, they're objecting to the hokey way "morals of the story" are presumably delivered. But a morality tale need not be hokey, nor must it suggest that our ethical dilemmas are easy. For example, HBO's *The Wire* seems to be the leading candidate among critics for the greatest television show of all time. But it's also a deeply moralistic cry of sorrow and rage about the way America's drug war has corrupted the fighters on both sides.

And while antihero shows like *Breaking Bad* and *The Sopranos* are praised for their sophisticated storytelling, the ethical messages are as stark and clear as any morality play's. Thanks to the painful ambiguity of the final episode of *The Sopranos*, we will never know what happened to Tony Soprano. But one thing's pretty clear: Tony is a bad man who will not live happily ever after. He may well survive his supper, but he will live out the time left to him in a condition of dread. Similarly, *Breaking Bad* ended with the drug lord née school teacher Walter White lying gut shot on a concrete floor as Badfinger's 1971 hit "Baby Blue" plays in the background. The first line of the song: "Guess I got what I deserved."

The second reason many tellers and critics think they are against "morals of the story" is itself deeply moralistic. When it comes to the ethical character of stories, history's tetchiest scolds have mainly been conservative types (Plato in *The Republic*, the Elizabethan critic Thomas Rymer, Anthony Comstock at the turn of the twentieth century, and so on) who've advocated for a harsh, narrow, and old-fashioned morality. Liberal storytellers may decry this closed-off morality, but it doesn't mean that they aren't themselves fiery moralists. They're just, as with the writers of *The Wire*, promoting complex, multilayered moral views, not stodgy conservative ones. They mistake what they object to: they think they're against "morals of the story," whereas they are really opposed to conservative morality.

But if most stories are supposed to be either explicit or implicit morality tales, then what about Hitler's highly dramatized life story, *Mein Kampf*? Was that moral? What about Thomas Dixon's nineteenth-century novel *The Clansman*, which did so much to popularize and spread the mythology of the Ku Klux Klan and inspired

The Birth of a Nation? Was that a moral story? In both cases, of course, the answer is a resounding *no!*

And yes.

The tireless moralism of storytelling has a very dark, dark side, a point I'll cover in depth in the next two chapters. For now, the claim I'm making isn't that it's impossible to write wicked stories or that all stories contain a moral message that everyone can agree on. For example, a story about the depravity of abortion would split modern American audiences down the middle.

My claim is narrower: the moral structure of prosocial protagonists holding the group together against antisocial antagonists applies as an overwhelming statistical reality. So *Mein Kampf* is that story (Jews and other undesirables are the archvillains of civilization and must therefore be eradicated). *The Clansman* is that story (black people are an existential threat to the dominant culture and must therefore be ruthlessly suppressed).

To sum up, I'm saying that stories are generally less moral—in the sense of capturing universal principles—than they are moralistic. In the same way that it's hard to write a compelling story that lacks a thorny problem, it's very hard for tellers to escape the deep moral gravity of stories. Problem structure and moralistic structure are the twin stars that tale-tellers helplessly orbit around. It's possible, with exertion, to bust free of this orbit, and some tellers have tried. But they've mostly found that few people want to follow them as they break out of the comforting groove of the universal grammar and float off into the cold, black void.

The deeply moralistic, judgy character of stories is embedded in the very word *story*, which is derived from the ancient Greek *historía*. This is obviously where we get our word *history*, but it's also where we get our word *story*. The oldest meaning of the root word *hístōr*, going back to how it's used in Homeric-era Greek, indicates a referee, wise man, or judge.[40] This suggests that stories, including historical stories, aren't just neutral accounts of events but renderings of judgment upon them.

The judgy connotations of the word *story* have died away along with knowledge of ancient Greek. But the judginess of our stories is nonetheless as pronounced as ever. A story, as the media psychologist Dolf Zillmann puts it, turns its consumer into a "moral monitor who applauds or condemns the intentions and actions of the characters."[41] And we play our monitoring role with gusto. We love the sensation of righteous indignation and the satisfying payoff of justice delivered. As the literary scholar Northrop Frye points out in *The Anatomy of Criticism*, "In the melodrama of the brutal thriller we come as close as it's normally possible for art to come to the pure self-righteousness of the lynching mob."[42] And studies back this up: people get more satisfaction out of stories in which offenders are punished rather than forgiven.[43]

In this chapter, I've argued that the unstoppable moralism of stories has a big upside for within-group bonding. In the next chapter, I focus on the downside of between-group division. The universal grammar of stories is paranoid and vindictive. Stories show us problem-drenched worlds and encourage us to turn on the people who are lousing things up. In other words, to proliferate narratives is to proliferate villains. To proliferate villains is to proliferate rage, judgment, and division between groups.

To make this case most effectively, I must first address what many experts say is the grandest thing about story: its role as an empathy generator.

5

THINGS FALL APART

Turning and turning in the widening gyre
The falcon cannot hear the falconer;
Things fall apart; the centre cannot hold

—William Butler Yeats, "The Second Coming"

June 1994 was one of the bloodiest months in human history. Rwandan Hutus suddenly rose up and began slaughtering their Tutsi neighbors and coworkers. The shocking ability of the Nazis to murder six million Jews over the course of World War II is usually chalked up to famed German organization, efficiency, and mechanization. But the Hutu massacre of the Tutsis was far more time-efficient despite the medieval nature of most of the weaponry—knives, screwdrivers, some guns, nail-studded bats called *masu,* and especially machetes.[1]

This was no pushbutton genocide; it was a genocide by close-range hacking and stabbing. The Nazi machine was as clean and antiseptic as they could make it. In the death camps, Nazis could kill without getting blood on their hands. The Rwandan *génocidaires* had to bathe in it. You have to *really* want people dead to kill in this way. Over four months, the Hutu death squads killed 333 Tutsi per hour—three times the rate of the Nazi death factories. But most of the approximately 800,000 victims died in the month of June.

Ten years later, families and friends across Rwanda huddled around radios listening to a popular soap opera called *New Dawn* while a team of social scientists studied how it affected them.[2] *New Dawn* was specifically designed to promote healing. It was in most ways an ordinary soap opera, but its creators—a Dutch NGO working with local Rwandan talent—wove themes of reconciliation into a Shakespearian plot featuring a young couple who forgo the logic of blood feud and taboos against interethnic marriage in order to forge a life together.

The researchers followed a large and ethnically diverse sample of Rwandans as they listened to the show, got hooked on it, and discussed it avidly in their free time. Listeners also filled out surveys about their social attitudes before hearing the show, and then again at the end of the study.

The show wasn't a panacea. It couldn't bring back dead loved ones or erase bitter memories. But *New Dawn* performed the empathetic magic of story: it transported listeners not just into a different world but into the skins of "the other." The show helped listeners appreciate that people from the rival ethnic group were precisely like them. And at the end of the study listeners were more likely to approve of interethnic marriage, to report higher levels of interethnic trust, and to endorse the necessity of working out differences through conversation not violence.

Stories are empathy machines.[3] When they really work, we're transported into different worlds, into different minds. Stories help us de-otherize one another in the most extreme way imaginable—"they" become "us." At their best, stories show us that our differences are mirages, our prejudices unfounded. To cite just one study, children who read the Harry Potter novels absorb not just

J. K. Rowling's ripping good tales but also her personal attitude of tolerance. The children show a reduction in negative attitudes toward marginalized "others" like racial minorities and gay people.[4] (This is true even if Rowling's Twitter feed has recently triggered debates about her own tolerance of transgendered people.)

In her book *Inventing Human Rights*, the historian Lynn Hunt argues that the so-called human rights revolution of the late 1700s—in which ancient human practices like slavery, patriarchy, and judicial torture came under sudden and sustained assault—was largely driven by the rise of a new form of storytelling: the novel. Unlike the theater, the novel gave audience members the illusion of immediate, transparent access to not just a person's exterior words or actions but also to their inner thoughts and feelings. According to Hunt, novels taught people to empathize outside of their family, clan, nation, and gender, and in so doing, they helped catalyze the most important moral revolution in human history.

Hunt's case is based largely on a suggestive correlation of the rise of the novel and the birth of the concept of universal human rights. But the plausibility of her claims has since been greatly enhanced by concentrated research on the empathogenic effects of stories. Studies show that a story like *Uncle Tom's Cabin* doesn't just lead white readers to feel more empathy for black people. They suggest that empathy is a kind of muscle, and the more exercise we give it by consuming fiction, the stronger it gets. Well-publicized studies have shown that heavy fiction consumption correlates with higher scores on tests of empathic capacity. These results seem to hold even when researchers control for the possibility that people who already have high empathy might naturally gravitate to fiction.[5]

For all these reasons, stories are widely celebrated by great artists, thought leaders, and scientists as our best hope for breaking down prejudice and tribalism and encouraging us to behave more humanely to more types of humans. But maybe there's a qualm growing fast in the back of your brain. That qualm is called "bad theory-data fit." Now that we have so much more story than ever before, has our empathy increased apace? Do we seem better able to understand each other across the old divides of politics, race, class, gender, and so on? Does it seem like we're doing a better job of the whole "with malice toward none, with charity for all" thing? Or does our story-glutted age feel at least as fractious and callous as ever?

So the authorities have some explaining to do. If stories have such balmy effects, why has the big bang of storytelling not coincided with a big bang of harmony and empathy?

Empathetic Sadism

"Art," as the novelist John Gardner puts it in his lovely book *On Moral Fiction*, "is one of civilization's chief defenses, the hammer that tries to keep the trolls in their place."[6] He's talking here about art generically, but the whole focus of his book is on story art specifically. Gardner is expressing a common and ancient sentiment in a powerful way. And he's both dead right and deadly wrong. Like most people, Gardner's love of story art makes it hard for him to fully appreciate that the trolls are armed with exactly the same narrative hammer and it's their primary tool for smashing worlds apart.

In fact, wherever you find capital *E* corporate-scale Evil in the world (as opposed to garden-variety mugging and murdering), you'll always find a story at the bottom. Remember that law of history laid out in the introduction: *Monsters behave like monsters all the time. But to get good people to behave monstrously, you must first tell them a story—a big lie, a dark conspiracy, an all-encompassing political or religious mythology.* You have to give them the kind of magical fiction that turns a bad thing—like hacking hundreds of Tutsi to death as they cry out for mercy in a church—into a good thing.

In propaganda stories spread on the radio, in newspapers, and on television, the architects of a supremacist Hutu mythology depicted the Tutsi as a kind of verminous invasive species that, if not eradicated at once, would surely rise up and destroy the Hutu. Both the genocide and any healing resulting from the soap opera *New Dawn* resulted from stories snapping through the thin membrane between the real world and storyland. But by any measure, the "Hutu Power" stories of hate and division were more powerful than *New Dawn*'s stories of love and reconciliation. The former blew through the membrane in a civilization-crashing deluge, whereas the latter came through in a just-detectable spritz.[7]

But scholars and journalists have celebrated *New Dawn*'s spritz while saying little about story's pivotal role in the great deluge. This is in keeping with a larger pattern in which people praise story generally, and story-generated empathy in particular, while genuinely seeming not to notice that story is a mercenary that sells itself as eagerly to the bad guys as the good.

The mythology of Hutu Power didn't take hold, and couldn't have, by being anti-empathetic. In fact, the myths generated enormous empathy among the Hutu—just not for the travails of the Tutsi "other" but for the suffering and humiliation of the Hutu in-group. As the psychologist Paul Bloom argues, empathy isn't always a good thing: "It is far easier to empathize with those close to us, those who are similar to us, and those we see as more attractive or vulnerable and less scary. Intellectually a white American might believe that a black person matters just as much as a white person, but she will typically find it a lot easier to empathize with the plight of the latter than the former. In this regard, empathy distorts our moral judgments in pretty much the same way that prejudice does."[8] In other words, because it's so much easier to empathize with in-groupers than out-groupers, the main effect of story-generated empathy may not be blurring us–them lines but powerfully reinforcing them.

We think that people who commit the worst kinds of violence are low-empathy psychos. Sometimes they are, but not always. The suicide bomber, for example, goes to his death drenched in empathy.

It's his powerful empathetic connection to the suffering and deprivations of his own people that drives and justifies his punishment of the enemy. The suicide bomber hates so much partly because he loves so much. And all that hate and love has been driven into him by stories—by true history, by ancient religious myths, and by his deep immersion in tales of evil conspiracies.[9]

So the big bang of storytelling actually *has* led to a big bang in empathy. But that empathy, to say the least, isn't always expressed in the ways we'd like. To the extent that stories divide people into good and bad categories, they generate a unit of callousness for every unit of empathy. Stories, in the act of generating empathy, also generate the opposite of empathy: a kind of moral blindness to the humanity of whoever is forced into the villain's role.

If a story is good, we become all fused up with the protagonists via mechanisms of identification, and we may fall in love with them a little or a lot. But we also have a different type of empathy-dependent energy circulating very hard and fast as we live through stories. This energy is hate. We hate the villain who's inflicting pain on the protagonist—pain that, thanks to the mechanisms of empathy-based identification, actually hurts us too. To facilitate our ability to experience the exquisite pleasures of hate, classic antagonists are generally drawn as flat, simple, and unchanging. And it's the flattened out, generic wickedness of antagonists that helps us transfer the hate we feel for them to the group they seem to represent—whether it's the stock villain of the frat bro, the inner-city gang kid, the vamp woman, the Wall Street banker, or the Tutsi fat cat.

"Conflict is the fundamental element of fiction," writes Janet Burroway in her classic creative writing guide *Writing Fiction*. "In life, conflict often carries negative connotations, yet in fiction, be it comic or tragic, dramatic conflict is fundamental because in literature only trouble is interesting."[10] This basic wisdom has been repeated so often, by so many authorities, with such unshakable conviction, that it is probably storytelling's first (and least negotiable) commandment.

But what sort of conflict do storytellers build around? Sometimes the conflict is between characters and the forces of nature, as in films like *Castaway* and *Mars*, or literary works like Jack London's classic short story "To Build a Fire" or the first part of *Robinson Crusoe.* I can even think of at least one story—the film *127 Hours* starring James Franco—in which a man literally plunges into a Vonnegutian hole and spends the whole film trying to get out.

But it's obvious that storytellers are fixated mainly on social conflict. The "trouble" Burroway references is driven overwhelmingly by conflicts among people. The more intense the clash of wills and wants, the more we're enthralled. That people gravitate most naturally to tales of social conflict is supported not just by the relative prevalence of these stories but also by research showing that even little children are far more attracted to stories of social conflict as opposed to other kinds.[11]

All of this is so familiar that it's easy to take for granted. But storytelling's intense focus on social conflict helps shape some of the most prominent, and least lovely, facets of human nature. Classically told stories divide the world into "us" (the world of the protagonists) and "them" (the world of the antagonists). More than that, they define the "us" mainly in contrast to the villainous "them." The villain is the other. The villains are "bad" and they deserve the horrible or degrading things that usually befall them in the end. Stories, as Fritz Breithaupt writes in his 2019 book *The Dark Sides of Empathy,* evoke "empathetic sadism," which he defines as "the emotional and intellectual enjoyment that most people feel in situations of altruistic punishment,"[12] for example, when the good guy kills, captures, or humiliates the bad.

In-Group Amity, Out-Group Enmity

Earlier I argued that the human appetite for the universal grammar of stories emerged partly as an adaptation to the primordial challenges of group life. Stories saturated their tellers and hearers in the same tribe-binding norms. They endlessly admonished us to be princesses, not tigers—social weavers, not cleavers. And they taught us, through equally relentless repetition of the theme of poetic justice, that it pays better to be a cooperative good guy than a selfish bad guy. Human groups with strong fantasies that bound them together into well-functioning collectives would have outcompeted human groups that lacked them. And we, the grandchildren of these ancient storytellers, have inherited the earth.

But what if our ancient storytelling instincts can't keep pace with fast changes in the modern world? Our narrative psychology evolved when our ancestors lived in tiny communities where everyone was united by kinship, language, ethnicity, and the same stories of cultural identity.

This small-scale world is all but gone, supplanted by nations containing dizzying multitudes of unrelated people. But the stories we tell are still doing their ancient work of carving people into tribes and setting them against each other. In the worlds of our ancestors, it may have made real sense to demonize the people on the other side of the river. They were potentially dangerous competitors. And the more a tribe imagines itself to be surrounded by bad guys, the more powerfully it clings to itself and holds itself together.[13] (The historian Carl Deutsch wryly encapsulates this unhappy reality: "A nation is a group of people united by a mistaken view of the past and a hatred of their neighbors.")[14]

But in multicultural and multiethnic societies, the tendency for story to serve as a tool not only of tribal formation but also of tribal division remains undiminished. And, allowed to run its course, it leads to intense tribal conflicts *within* a society, leading plausibly to outcomes like cultural balkanization and even civil war. To keep this discussion from getting too fuzzy and abstract, let's look at a specific genre, historical narrative, as an example of the way story can facilitate in-group amity and out-group enmity.

"The Sores of History"

At the outset of his book about the universal human need to storify the past, the philosopher Alex Rosenberg sets out to "show that historical narratives are wrong." Which historical narratives? the reader wonders. "All of them," answers Rosenberg.[15]

Rosenberg doesn't deny that we know things about the past. Thanks to historians, we know when the pyramids were built and a bit about how. We know when the Declaration of Independence was signed and how it was composed. We know when World War II started and how the state of Israel was formed. We have libraries full of trusty historical data about names, places, dates, and events. But, Rosenberg maintains, history turns shaky when we weave this data into stories explaining why things happened as they did and what it all means. And the histories we all carry around in our heads, which are made up of dimly remembered school lessons and all the movies we've seen and hazy recollection of books and documentaries and some stuff told to us by some guy at some bar—what might be called our collective historical memories[16]—are most suspect of all.

To this point, Rosenberg sounds like a provocative guest trying to enliven a boring dinner party. But he has a bigger point: all historical storytelling is to a large degree not only wrong but dangerously so. In this assertion, Rosenberg is attacking George Santayana's old saw, "Those who can't remember the past are doomed to repeat it."[17] This cliché, which most of us accept without resistance, allows historical storytellers to believe they're weaving true stories that will help us avoid the catastrophes of our gory past. For Rosenberg, it's remembering (and misremembering) our past that's more likely to bring on the gore.

Similarly, the journalist and essayist David Rieff provides many examples where rival historical stories—which often come down to bickering over who did what to whom—drive cycles of immiserating conflict. Reciting these "conflicting martyrologies"[18] keeps "the sores of history"[19] from scabbing and healing. After describing the clash of histories driving conflict in places like the Middle East and the Balkans, Rieff concludes with a radical proposal: "The overall point of this book is that sometimes—maybe more often than not—amnesia is better than memory."[20]

History, as an important genre of storytelling, is heir to all of the structural patterns we find in story more generally, with all the wholesome and noxious effects. As we saw in Chapter 4, historians give us a focus on negativity that's as intense as that of fiction or journalism. They give us sharp, clear focus on conflict and trouble, and they frequently give us a full-blown moral grammar of villains menacing victims and heroes racing to save the day.

Of course, not all histories give us such sharp and rousing grammar, and these are precisely the histories that Rosenberg and Rieff endorse. These tend to be academic accounts that lay out all the clutter of historical facts, while resisting the urge to neatly weave true facts into merely truthy tales. But as with other forms of storytelling, the histories that are most successful, that climb bestseller lists, and that compete most effectively for space in our societal memory banks are those that most fully impose the universal story grammar on the disorder of the past: they give us good guys locked in struggle with bad guys, with life-and-death stakes. They also tend to be intensely judgy affairs, where, if the good guys don't win, at least we have the satisfaction of condemning the bad.

For this reason, Rieff and Rosenberg both see history as a form of historical fiction that often draws groups together in opposition to one another. As usual, Rosenberg pulls no punches: "Stories historians tell are deeply implicated in more misery and death than probably any other aspect of human culture. And, as we'll see, it's the nature of the most compelling stories they tell that's responsible for the trail of tears, pain, suffering, carnage, and sometimes extermination that make up most of human history."[21] He points out that, for centuries, "ethnic, linguistic, and religious groups have been mobilized by their proprietary narrative histories of mistreatment, discrimination, even genocidal suppression into the mistreatment, discrimination, and genocidal suppression of other ethnic, linguistic, and religious groups."[22]

In the famous interrogation scene of George Orwell's *1984*, the confessor/inquisitor O'Brien makes our hero, Winston Smith, recite a party slogan: "Who controls the past controls the future: who controls the present controls the past." Regardless of whether history is capable of capturing truth, the slogan suggests, it's always about capturing power. It's a Big Brotherish way of saying "the storyteller rules the world." And to the extent that this aphorism is true, our debates about history are as much about competing for future-tense power as for defining past-tense fact.

If the past given to us by historical storytellers is, as the saying goes, a foreign country, the country is an absurd, psychedelic place where nothing is stable and meaning is in perpetual flux. Of course, this is partly because new evidence dug up by archivists or archaeologists is always coming to light, enabling us to see the past in a more complete way. But it's also because history is a fun-house mirror of the ever-changing present.

Narrative history, I propose, can be defined as the imposition of the imagination of the present on the defenseless corpse of the past. History is a shaping and editing and massaging of the unruly past in order to create neatened-up stories that serve the needs of the present. History is therefore a self-portrait of "us"—our concerns, our obsessions, our grievances, our power struggles—perhaps even more than a portrait of "them." To sharpen this point a little, it's conventional to argue that science fiction projects our current obsessions onto the future. But we can also see history as a genre of speculative narrative that projects our current obsessions onto the past.

Noble Lies and Ignoble Truths

In the beginning, Plato tells us in *The Republic*, the gods grew humankind from the soils of Mother Earth and they sprouted forth as brothers and sisters. In those destined to rule, gold was mixed with the soil. In those destined to serve as soldiers, silver was mixed in. And for the common sort, fated to sweat in the fields or the kitchens, brass strengthened their bones. This is why rich men never blister their hands on the plow. And this is why the sons of poor men, weighed down by baser metal, never rise to rule.[23]

But, in a second myth of his own invention, Plato explains that lowly men can rise to greatness after death.[24] A warrior named Er is cut down in battle, then lies dead in the grass for ten days before waking again to life. Er tells his comrades that he descended to Hades, where he saw the logic of all existence laid out like the hidden joinery of some great ship. In Hades, the shades of good people traveled up to heaven, where they knew a thousand years of unearthly delight. Bad people went down to a hellish realm where they endured a thousand years of gruesome punishment. Then, all the good and sinful souls gathered in a field to await reincarnation. Each person could be reborn exactly as they wanted—as an animal, as an honest craftsman, or as a tyrant who would luxuriate in the delights of his power and his flesh.

Each of these two myths is, in Plato's famous phrase, a "grand and noble lie."[25] What matters most about a story, Plato asks: If it's true or if it does good? If a fiction steers people toward virtue and makes them treat one another as brothers, Plato concludes, the lie would be nobler than the truth.

Plato set out to draw a blueprint for a future utopia founded on the bedrock of unshakable reason. But what a nice coincidence for Plato that the blueprint would give aristocrats like him a forever monopoly on power. Plato's origin myth naturalizes the traditional power structure. And the Myth of Er tells unfortunate people not to blame their fate on anyone else or on what we'd now call the "structural" imbalances in society. The Myth of Er is that every person—privileged and unprivileged—has received exactly the life they requested in the underworld. Together the two myths make the caste system not just natural but fair.

Maybe every great nation is founded on similar noble lies. The great African American writer James Baldwin neatly describes America's founding myths from the position of a man, and a community, who was not well served by them:

> The American Negro has the great advantage of having never believed the collection of myths to which white Americans cling: that their ancestors were all freedom-loving heroes, that they were born in the greatest country the world has ever seen, or that Americans are invincible in battle and wise in peace, that Americans have always dealt honorably with Mexicans and Indians and all other neighbors or inferiors, that American men are the world's most direct and virile, that American women are pure. . . . Negroes know far more about white Americans than that; it can almost be said, in fact, that they know about white Americans what parents—or, anyway, mothers—know about their children, and that they very often regard white Americans that way. And perhaps this attitude, held in spite of what they know and have endured, helps to explain why Negroes, on the whole, and until lately, have allowed themselves to feel so little hatred. The tendency has really been, insofar as this was possible, to dismiss white people as the slightly mad victims of their own brainwashing.[26]

Baldwin is describing the doctrine of American exceptionalism. This is the idea that America is history's chosen nation; that we're a collection of self-selected explorers, pioneers, and wild-eyed idealists bringing the flame of enlightenment and freedom to a benighted world; that we are, to put it humbly as a true American should, the triumphant apotheosis of the human struggle for liberty and self-determination.

This complex of semi-fictions, which I will refer to as Myth I, can seem pretty silly now. But we spent hundreds of years building this noble lie about ourselves in order to sew together a sprawling swath of fractious states and territories and religions and ethnicities into something like a single national entity.

And then, after many generations spent building the lie and genuflecting in front of it, 1960s-era intellectuals and activists tore it down. They did so not just because it was riddled with falsehoods and glaring omissions but also because they recognized the lie as nakedly political. Myth I was, like Plato's noble lie, power politics disguised as history. As is now well understood, the lie left out the lesser status assigned to women and minorities of all kinds. It downplayed the demented cruelty of the slave trade and the continuation of black immiseration that followed. And it was all but silent on the near-annihilation of indigenous people on the North American continent.

Sixties academics and activists tore down America's noble lie while building up a new counterhistory of ignoble truths, which I will refer to as Myth II. Under Myth II, which is now the dominant story on the American Left, the sun-drenched annals of American history flipped over suddenly, sickeningly and became a history from the point of view of the enchained, the plundered, and all the restless ghosts of the murdered. In some telling of the ignoble truth, America—and the Europe from which it rose—became a monster civilization that was all the more hideous because of its unshakable faith in its own special goodness. Behind almost everything sinister in the world was found the cheeseburger-grease fingerprints of the American colossus. The noble lie (Myth I) painted America as the shining city on the hill. The ignoble truth (Myth II) painted it as the heart of Mordor—and that shine from the hilltop was the Eye of Sauron.

The transition from "noble lies" to "ignoble truths" was a process of simple inversion: the white man went from *bearing* the world's burdens to *being* the world's burden. It took the rose-colored glasses previously used to observe America and simply deepened the color of the lenses until all that could be seen was blood. The white man switched from hero to villain, but never stopped being a caricature. Myth II is still a story of American exceptionalism—but with America being exceptional only for its brilliance at self-delusion, oppression, and greed-fueled destruction.

For two reasons neither grand historical narrative is true—not Myth I's noble lies nor Myth II's ignoble truths. The first reason is concrete. Myth II can be seen as a prosecutor's one-sided case against Western civilization since the Age of Discovery, and Myth I can be seen as the equally one-sided brief for the defense. But they're both distortions forged from edited history. And what has been edited out of one is what's magnified in the other. If you shuffle the two stories into each other like two incomplete stacks of playing cards, you get something like complete history. The second reason these myths are untrue is a little more abstract. Neither myth is true because they're both stories, and stories are never quite true.

Here's a hard but important thing to try to wrap one's mind around: *No story ever really happened.* Life happened. Shit happened as people tried to get by. But no story has ever happened in the present tense. A story is *always* an artificial, post-hoc fabrication with dubious correspondence to the past. So, whenever we encounter a highly grammatical history of villains and heroes, evil and good, we should be on guard. Our minds are designed to deal with complex reality through narrative simplification. We do this by forcing the universal grammar down on experience like a story-shaped mold. And the mold transforms the messy past into a neatened-up historical fiction.

Battling Books

Amy Chua's *Day of Empire* (2007) is about the rise and fall of the relatively small number of nations in world history that were unrivaled in military, economic, and cultural realms. So-called hyperpowers—the Mongols, the ancient Roman Empire, the British Empire at its height, and the United States since World War II—gained supremacy by being remarkably tolerant societies. It's important to say that all hyperpowers were awfully discriminatory in plenty of awful ways, just not compared to competitor nations. As a result, they were able to absorb the human capital—the genius and the sheer muscle—of diverse cultures and ethnic groups.

A pretty story so far. Greater tolerance of diversity is a sine qua non for the rise of hyperpowers. But it's also key to their eventual declines and falls. Again and again, Chua describes a historical trajectory whereby hyperpowers thrive through relative tolerance but then hit a tipping point when the glue of national identity can no longer hold and the demographic crazy quilt frays apart at its identity seams.

A hyperpower is a wild collage of different types of people. They aren't held together by natural barriers of geography or phantom lines drawn on maps or shared blood. They're held together by stories—myths, folktales, popular culture—that reinforce shared values and a sense of common identity.

As Chua sees it, America has been, like other hyperpowers, remarkable for its tolerance of diversity. Despite its struggles with nativism and racism, it has also welcomed an astonishing range of immigrants from all around the world, and it hews to a definition of Americanness based not on blood and soil but on a shared set of ideals laid out in its founding documents. Since passage of the revolutionary civil rights bills of the 1960s, Chua argues, America has finally begun "however fitfully and imperfectly, to develop into one of the most ethnically and racially open societies in world history."[27]

I thought of Chua's book when, in February of 2020, I watched the forty-fifth president give an address to a joint session of Congress announcing, as presidents do, that "the state of our Union is strong." But the spectacle itself revealed worrying weakness. It began with an impeached and as yet unacquitted president refusing to shake the hand of the Speaker of the House. And it ended with Speaker Nancy Pelosi standing on her perch above the president's head and theatrically shredding his printed speech, page by page. But what worried me most wasn't these displays of mutual contempt but the reactions of the lawmakers to the president's remarks. The Republicans sitting to one side were rapturous in their applause, chanting "Four more years!" and "USA!" with expressions of unfakeable euphoria on their faces. The Democrats sitting to the other side squirmed in their chairs with expressions toggling between agonized disbelief and just plain agony.

We live in a representative democracy, which means the people in that room represent us. They represent us in the sense of doing our political bidding. And they represent us, in a more abstract sense, as a microcosm of our anger, our dysfunction, and our just-suppressed violence.

What I saw that night was chilling proof of the two Americas some politicians say doesn't exist. And the room was divided not only down party lines but down identity lines. It was impossible not to notice that the president was addressing himself almost exclusively to the side of the room consisting overwhelmingly of gloating white males. The people he mostly wouldn't look at also included plenty of white males, but plenty of everything else too. I went to bed in fear. I felt I'd seen the conditions that eventually culminate in civil war or, if that seems too hysterical, at least an endless cold civil war that freezes progress and guarantees steep national decline.

The first impeachment, the State of the Union Address, and everything else were soon forgotten as the Covid-19 pandemic accelerated and, a few months later, the nation was convulsed by the grisly killing of George Floyd by Minneapolis police officers. Scenes of great beauty ensued: people of all races hit the streets to march for racial justice and equality. And, for a time, there seemed to be an emerging transpartisan will to do *something* about racial disparities in America. But there was ugliness, too: vandalism, looting, and, most disturbingly, running street battles between extremists on the right and left.

In Jonathan Swift's "The Battle of the Books" (1704), the volumes in a library literally come to life, and two armies of books—one representing the intellectual giants of Greece and Rome, and the other representing modern thinkers—go to war over their competing notions and narratives. It might seem like a far-fetched metaphor, except that virtually all grand-scale human conflict goes back to battles of narratives, whether written in books or not.

The rival groups of protesters—as well as the legislators listening to the president's speech—were playing out a battle of the books. Most of the left-wing protesters and legislators lived somewhere in the story of Myth II. And most of the right-wing protesters and legislators—and certainly the most ardent inhabitants of MAGA land—lived somewhere in the story of Myth I.

With all her tribes and factions mustering online and in the streets to roar their resentments, America looks like a hyperpower losing the common story that once pasted it together. It would be dumb to panic. Our nation has been through worse times and emerged stronger. But it would be equally dumb to yield to complacency. As Amy Chua points out, great civilizations are so far batting a thousand when it comes to blowing it, and just because we haven't definitively blown it yet doesn't mean we aren't in the process of doing so.

What's testing our national glue more than anything else isn't just our troubled history but the incompatible stories we tell about it. If we're going to bring the national house back together, we have to learn to tell a story capable of inspiring us to move forward as protagonists on the same quest.

I don't know exactly how we can do this. But I do have a promising idea for how we might make a start.

History Without Villains

We can imagine story as a floating abstract presence that seeks out disorderly information and organizes it in predictable ways. This organization of information has powerful effects on human beings. It activates us. It rallies us. It binds tribes together and keeps them bound. But story also distorts reality because tellers always have powerful incentives to wrench real-world facts into line with the most powerful grammar of fiction. This is as true for gossip spreading through a clique as it is for novels as it is for the stories of American history that rise off the page to swarm through our imaginations.

When it comes to history, the grammar isolates problems in the modern world and then jumps back into history to indict and sentence the agents responsible for our woes. Historical storytelling—not just in America but everywhere—frequently amounts to a kind of revenge fantasy, where the malefactors of our past can be resurrected, tried, and convicted for violating moral codes they frequently hadn't heard of.

Story emerged as a tool of tribal cohesion *and* competition. It cohered tribes so they could compete more effectively with other tribes. This served us well in the monocultural societies of the distant past, but in today's multicultural societies it fuses tribes *within* the nation and pushes them apart from others. Our historical myths have always united by dividing, and they will continue to do so unless we fundamentally alter how we narrate the past.

Here's where hope lies. The great majority of tales ever told cross-culturally and cross-historically hew to the universal grammar. In particular, successful stories (in the specific sense of attracting and delighting broad audiences) that *don't* highlight conflict and trouble are vanishingly rare. But when it comes to the moralism of tale-telling, countless exceptions show that, although stories usually appeal to the judgier angels of our nature, they don't have to. This is a good sign that we can tell stories about ourselves that are less divisive, less tribalistic, and less addicted to defining the "us" in contrast to a villainous "them"—if we can only summon the will.

Consider the intricate plot of the film *Babel*, which spools out in tragedy from beginning to end. An American couple loses their infant to SIDS, then the wife is shot, and their two small children wander lost in a sweltering desert. The American woman is shot accidentally by two young Moroccan brothers, and this ultimately leads to the death of one of the boys and the destruction of their family. The two American children are lost in the desert as a result of a series of misjudgments by their loving undocumented nanny, who is then roughly deported to Mexico. The Moroccan boys had been playing recklessly with a rifle left behind as a gift by a Japanese hunter, and when the hunter's wife later commits suicide, his distraught daughter is pushed out on a ledge of her own.

Babel is slavishly true to the universal grammar's penchant for extreme conflict and trouble. From beginning to end, all the characters are coping with horrendous problems. But the judgy side of the universal grammar is utterly absent. There are no bad guys in *Babel*. The entire plot of this death-haunted film is driven by good people acting out of love. Small connections between people, seemingly insignificant moments of carelessness or kindness, cause butterfly effects of great suffering.

Rewatching *Babel*, I was struck by the artificiality of fictional villainy. Of course the protagonists of a film like *Babel* are artificial, too. But at least, as E. M. Forster famously put it in *Aspects of the Novel* (1927), there's an effort to make them "round." Protagonists may be nothing but collages of words, but they rise from the flat page like three-dimensional people. They have families and weirdo tics. They have dogs they love or resent. They have character flaws they want to overcome.

Villains in a novel or a screenplay are also just collages of words. But they usually don't rise off the page in quite the same way. More often they're flattened out in ways that deny their full humanity. Antagonists are frequently drawn as machines running appetite algorithms. And unlike protagonists, who usually experience some form of moral awakening (or at least moral evolution) in the course of their adventures, antagonists are typically morally static. Outside of true sociopaths, people this simple hardly exist in real life. They have to be fabricated by storytellers.

And here's another thing that struck me after rewatching *Babel*: maybe the villains swarming through our history stories are just as artificial as those in our fiction. And maybe we should try composing histories that do without the villains altogether, just like *Babel*.

Stripping out the villains seems like a plausible way of narrating history in the abstract but faces difficulties as soon as we turn to the hard cases. How can we narrate episodes of mass horror—like the Atlantic slave trade or the Holocaust or the genocide in Rwanda—without naming and shaming the villains? Wouldn't this be precisely the sort of noble lie Plato recommends—a lie that erases the past, rewrites it in a prettier way, and ultimately ends up extending current injustices forever into the future?

In short, wouldn't histories without villains be, in themselves, fairly villainous fabrications?

Empathy for the Devil

When I was a teenager, I remember driving to the grocery store with my dad. I don't remember the context now, but somehow our discussion turned to morality. My father said something that has stayed with me ever since. "I'm no better than a criminal," he said. "I'm going to go into the Price Chopper to buy a loaf of bread. But if I was poor, and my kids were getting thin, and my wife was desperate, I'd go into the store and steal the bread. And maybe I'd do worse. Maybe I'd sell drugs or break into houses. And if I did, people would call me a criminal—a bad person. But I have enough money to pay for the bread, so I get to be a good person. But I'm not a good person . . ." And here he paused as he searched his mind for the right words, "I just have the luxury of virtue."

Even then, as a rude and dumb boy, I was struck by the power of that phrase and the shock of the underlying idea: virtue isn't an inherent property of character so much as *a luxury product* that comfortable people can easily afford and others buy at much higher cost.*

I don't think my dad knew it, but he was giving the gist of a great discovery in the field of moral philosophy—a discovery on the same level as that of an important principle in physics or chemistry. This discovery is controversial, even among philosophers, because it seems to dissolve the foundation of moral responsibility and with it the warrant for holding each other accountable for bad behavior.

Here is the breakthrough: we generally proceed as though morality emerges from a person's inherent character, not their luck. But in separate but related papers, the philosophers Thomas Nagel (1979) and Bernard Williams (1981)[28] showed that whether or not you behave morally is as chance-dependent as a game of cards. My dad's middle-class status didn't just allow him to resist the temptation to steal; his comfort meant he felt no temptation at all. But if life had dealt him a bad set of cards—including the genetic cards that underlie all of our psychological traits—he would have felt much greater temptation to sin and much greater reason to give in.

**The word *villain* was not commonly used as a synonym for the bad guy of a story until the early 1800s. For centuries prior, according to the Online Etymology Dictionary, a villain was just a low-born person, a "peasant, farmer, commoner, churl, yokel." Over time, as part of a process C. S. Lewis calls "the moralization of status-words" (Lewis 1959, 21, 118), the word *villain* morphed into a general term of abuse and then into a synonym for "bad man." So, built into our basic vocabulary of storytelling is a morally obtuse assumption that bad guys will generally be of low birth. This change in vocabulary didn't follow an enlightened realization that "immoral behavior" was often driven by economic conditions beyond individual control. To the contrary, it reflects an assumption that poor people are less moral due to the inherent poverty of their bloodlines, not their wallets.

If this seems less like a philosophical breakthrough and more like a statement of the obvious, imagine hypothetical twin brothers living in Germany prior to World War II.[29] The young men are virtually identical regarding all factors underlying the nature and nurture of moral behavior. But one brother moves to America to seek work some years before the Nazis rise to power. The other brother stays put. When World War II breaks out, the twins fight for their respective sides. And so one brother goes down in history as a despicable storm trooper and the other goes down as a hero of America's Greatest Generation.

But how can we condemn or celebrate either brother? Each brother's choices were almost entirely determined by his circumstances, not some inherent moral compass. If the American brother had happened to stay in Germany, circumstances would have put him on the wrong side of history. If the brother who stayed in Germany had instead followed his twin to America, he might have gone down as a hero too.

For us to classify a behavior as moral or immoral, a person must have a real choice. If someone overpowers you, forces a gun into your hand, and makes you pull the trigger, no one will hold you to blame. And it can be reasonably argued that the twin brothers also had no real choice in the matter and that to condemn one while exalting the other is as morally obtuse as it is emotionally satisfying. Neither brother deserves his praise or blame.

Many would protest that the German twin should have resisted. He should have seen the evil of the Nazi regime. But how could he? He was living inside a fascist version of *The Truman Show* and all of his inputs told him the same story. Many of us would like to imagine that, if we'd have been in the German brother's shoes, we'd have grasped the evil of the ideology, seen through the propaganda, and steeled ourselves against the rising tide of militarism and hate. If we're shameless enough, we may even fantasize that *we'd* have leapt up like heroes to resist before being knocked down as martyrs.

But this is sad vanity. Strikingly few Germans managed to resist, much less rise up against, the enthusiasms of the Nazi era.[30] They were narratively transported by an epic story of heroism and villainy that suspended their ability to disbelieve. And those whose disbelief couldn't be suspended were mostly terrified into compliance—they were up against Nazis, after all.

The most striking thing about the crimes of the Nazis, or any other instance of mass wickedness you can name, is they mostly weren't committed by the villainous flatties given to us by so many storytellers. They were committed by round people like us.

So, when it comes to slavery, in what sense can we claim moral superiority to our European ancestors who bought and used 12.5 million African slaves? In what sense are we better than our African ancestors who themselves used and abused many millions of enslaved people continuously over at least a millennium, while also exporting many millions more not just to the West but to all points of the compass?*

We can claim moral superiority only in this limited sense. We know now what our ancestors didn't: slavery is categorically wrong. But this moral edge is so thin, and it depends entirely on good luck. If you'd had the bad moral luck to be born German in the early part of the twentieth century, you'd probably have sided with the Nazis. If you had the moral bad luck to be born a white person in the South in the middle part of the nineteenth century, you'd likely have sided

* The common idea that Africa was simply a victim of the Atlantic slave trade is not supported by historians. Trevor Burnard sums up the scholarly consensus: "Recent research by historians has made clear that the key players in determining the shape of the Atlantic slave trade were African merchants and rulers. Europeans came to Africa upon sufferance and were dependent on local leaders for access to goods, including slaves. . . . The conduct of trade generally followed African dictates until European colonialism began in earnest from the late-nineteenth century, by which time the Atlantic slave trade had largely ended" (Burnard 2011, 83). In short, scholars argue that the Atlantic slave trade wasn't something Europeans did to Africans. It was something that powerful Africans and Europeans did to less-powerful Africans (see Gates 1999, 2010; Heuman and Burnard 2011; Mann 2011; and especially Thornton 1998).

with the Confederacy. If you were a strong man born in the western African kingdom of Dahomey in the eighteenth or nineteenth century, you'd likely have been a fearsome and ruthless raider of slaves, and you'd have glorified your king every year through the ritualized slaughter of hundreds or even thousands of captives.[31] Obviously, these instances could be extended to near-infinite length, with examples taken from virtually every culture throughout the bloody annals of human history.

The behavior of Nazis, Confederate white supremacists, and Dahomey warriors is villainous to us but was normal and in fact virtuous for them. They aren't worse people than us. They just had the moral misfortune of being born in cultures that, we now see, mistakenly defined bad as good. And if we had been born in such circumstances, we'd likely have behaved the same way.

Someday our descendants will look back and condemn even the most enlightened among us, not just for the sins we know about—for example, factory farming or the out-of-control carbon economy—but for the sins they think we *should* have known about. I propose they will be appalled by the way we made villains of each other, by the spectacular hypocrisy of our moral judgments. By the ways whites and blacks, blues and reds, believers and unbelievers, women and men appeared as caricatured villains in each other's morality plays. When we villainize, we dehumanize and give ourselves a free pass to sink into the voluptuousness of our sanctimony, if not our hate. And in so doing we make villains of ourselves.

This doesn't mean we shouldn't name the bad acts of our ancestors or that we can shuck off the duty of reparation. It means we shouldn't have the poor taste of confusing *our* moral good luck with moral virtue. That would be as obtuse as condemning the poor man who must steal his daily loaf and celebrating the well-fed man who doesn't have to.

What this way of thinking about the past requires, then, is empathy for the devil. We're encouraged to find empathy for the wretched of the earth—the weak, the poor, the enchained, the victimized. And the moral imperative of this isn't hard to grasp. It's contained in the eternal ethical wisdom: "There but for the grace of God go I."

But when it comes to the villains and victimizers of history, we have a failure of empathetic imagination. We won't allow ourselves to acknowledge that, when it comes to the slavers, inquisitors, conquistadores, and *génocidaires*, that's quite obviously where we'd have gone, too, if not for the grace of God. The devil isn't "the other"; the devil is us. He's who I would have been—who you would have been—if born into his circumstances.

6

THE END OF REALITY

The human world is made of stories, not people.
The people the stories use to tell themselves are not to be blamed.

—DAVID MITCHELL, *Ghostwritten*

IN 1944, THE PSYCHOLOGIST FRITZ HEIDER AND HIS RESEARCH assistant Marianne Simmel produced a very short, very crude animated film.[1] They cut geometrical figures out of cardboard and moved them around on transparent glass in the technique of stop-motion animation. The resulting silent film shows a small triangle, a big triangle, and a small circle moving busily around a rectangle. One side of the rectangle flaps open and closed, and sometimes the geometrical figures slide inside. At the end of the film, the small circle and the small triangle disappear offscreen, and the big triangle butts against the big rectangle until it breaks. Heider and Simmel showed the film to 114 research subjects, who were simply asked to describe what they saw. (Before reading further, please take ninety seconds to watch the Heider-Simmel film on YouTube.)

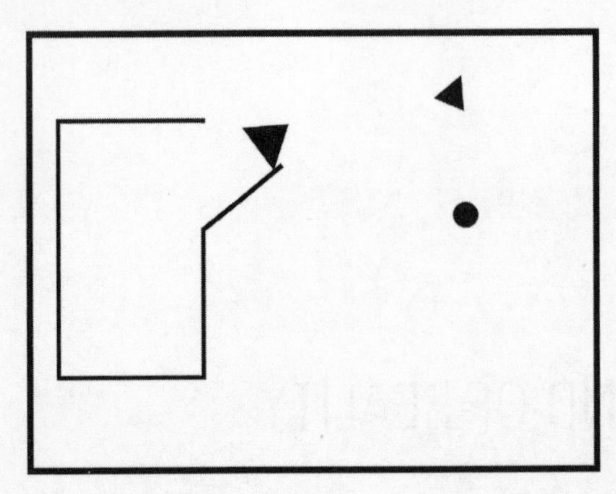

Re-creation of a screen shot from the Heider and Simmel film

I first saw the short film about sixty-five years after it was made and watched in delight as the simple geometry resolved itself into a classic three-act love story. Act One: The two lovers, Small Circle and Small Triangle, move side by side onto the screen. Act Two: Big Triangle decides he fancies Small Circle as well. He uses his pointy nose to wedge the lovers apart, then chases Small Circle into his house (the big flapping rectangle), where he tries to trap her in a corner. Act Three: To my relief, Small Circle slips past the lecherous Big Triangle and reunites with her mate outside. They race side by side around the house, with Big Triangle in hot pursuit. Finally, the lovers escape offscreen. Furious and frustrated, Big Triangle slams against the walls of his house until it collapses.

When I showed the film to my students, their reactions surprised and confused me. While many saw a love story like me, others were just as convinced that they'd seen a sordid family drama or a slapstick comedy along the lines of *The Three Stooges.* I should have been fascinated by the Rorschach element of the film, but I wasn't—not at first. I was too frustrated by my students' inability to see the "real" story that I *knew* Heider and Simmel were trying to tell.

At the time, I'd only heard enough about the Heider–Simmel film to look it up on YouTube, but I'd never read the original scientific paper. When I did, I realized to my embarrassment that my students hadn't been getting the film wrong; I had been. And it wasn't until years later that I recognized the film as a powerful illustration of the story paradox: it reveals something wonderful about our nature as storytelling animals while also hinting at something profoundly scary—something like the root of humanity's largest, deepest evils.

Although this wasn't divulged in the paper of 1944, Heider later explained in his autobiography that he did have a hazy story situation in mind when he dreamt up his famous film. "As I planned the action of the film," Heider wrote, "I thought of the small triangle and the circle as a pair of lovers or friends and I thought of the big triangle as a bully who intruded on them."[2] So rather than imagining a definite plot, Heider imagined an open framework consisting of a setup, a conflict, and a resolution, which could support a variety of basic plots, including Heider's visions of either a buddy flick or a romance.

Despite its built-in ambiguity, the film nonetheless produces impressive convergence in interpretation. In the original experiment, for example, 97 percent of the 114 people who viewed the film saw a story. Moreover, there were strong regularities in the stories people saw. First, even though the geometrical shapes look and move more like scurrying beetles, almost everyone automatically saw them not as bugs but as people in conflict. And most viewers automatically gendered the shapes in the same way—the circular female, the spear-headed males. On top of that, in a strong majority of cases people agreed on the basic protagonist–antagonist split: Big Triangle was the bad guy, and the two smaller figures were good. They also tended to give the shapes similar personality traits: Big Triangle was a bully, Small Circle was timid.

But there are equally impressive divergences in interpretation. For instance, although the most common story people see is indeed a love triangle, it's not always *my* love triangle. Some viewers were convinced that Big Triangle was the aggrieved party, having been cuckolded by Small Triangle. Others thought that Small Circle *wanted* to be with Big Triangle. He, not she, was the unwilling party. But more times than not, viewers saw no love story at all. Some viewers saw a tale of domestic violence in which Big Triangle was abusing his family. Others saw Big Triangle as a harmless oaf being harassed by Lilliputian invaders. One viewer thought Big Triangle was a witch, trying to catch two children.

Over the years, I've conducted scores of informal replications of the Heider–Simmel experiment in my classes and in public lectures, with total numbers of participants stretching to many thousands. And I came to love the way viewers project idiosyncratic stories and meanings onto such simple geometry. The diversity of response shows that when we see the film we're not *experiencing* story but are actually *creating* the story through a sequence of unstoppable cerebral reflexes.

Even when the film is run backward, and there can be no possibility that the psychologists had a definite plot or meaning in mind, people still see stories. When I watched the film backward, I expected to see my love triangle story running in reverse—like the bygone experience of hitting rewind on a VCR tape. Instead, I watched slack-jawed as my love story dissolved into a scene out of *A Clockwork Orange* in which the formerly predatory Big Triangle (now my stolid protagonist) is harried by a feral pair of anarcho-terrorists.

The point here isn't limited to how people interpret simple animation. The point is that this is what storytelling animals are up to all the time: we're trying to impose the meaningful and comforting order of story structure on the ambiguities of existence. And thanks to differences in our minds and experiences, we will *reliably fail* to see the same story—just like in the famous film.

Moreover, though our interpretations of a primitive cartoon couldn't matter much less, this effect runs riot through domains of experience where it couldn't matter much more. Especially because the stories we see can be divisive. When we experience chaotic events, we naturally construct stories to bring order to the chaos.

And, as with all of the most typical responses to the Heider–Simmel film, we're apt to resolve the chaos into a moralistic trinary of victims, villains, and heroes.

The story psychology revealed by the Heider–Simmel effect isn't the root of *all* evil. But this humble experiment digs to the root of the most tragic type: the kind otherwise good people get swept up in. I'm referring to the human tendency to glom on to a story often for no good reason at all, to cling to it tenaciously, to let it structure our worldview, and to allow it to project patterns on the world that aren't really there.

In all my screenings of the Heider–Simmel film, debates over the "true story" have often been raucous but always friendly and light-hearted. People don't get invested in their interpretations, not intellectually and not emotionally. But if you take the same narrative psychology laid bare by the film and you translate it to situations where (1) there's a truth that can conceivably be determined, and (2) the stakes run higher than cartoon interpretation, people will dig in for their fictions and fight.

Sometimes it seems to me that what I call the Heider–Simmel effect—our tendency to all watch the same film and see different stories—explains everything about the roiling anger and confusion of modern life. And this effect, souped up by technological and cultural upheaval, helps explain why we now find it so hard to converge on consensus narratives about the basic shape of reality.

You Don't Have a Narrative . . . Narrative Has *You*

The science I cover in this book contributes to a massive and ongoing reassessment of how people make decisions and change their minds. The old view is reflected in the name we gave to our species in a fit of colossal self-regard, *Homo sapiens*—wise person. According to this view, humanity's defining characteristic is our rationality. Our minds are designed to prioritize truthful conclusions based on careful evaluation of evidence. But why, then, are our reasoning faculties biased toward error in all sorts of predictable ways?[3]

In their 2017 book *The Enigma of Reason*, the psychologists Hugo Mercier and Dan Sperber ask what is an obvious question only in hindsight: What if we've mistaken the *reason* for human reason? What if we fault the sloppy engineering of our rational faculties because we're confused about what they are *for*?

Mercier and Sperber propose that rationality is a tool that evolved less to determine objective truth than to serve as a sword and shield for social competition—the sword that attacks in arguments and the shield that parries. From this point of view, an apparent *bug* in the brain, such as confirmation bias, is actually a well-functioning *feature* of the brain. And *Homo sapiens* becomes not so much a rational animal as a *rationalizing* animal. Rationalizations are fictions we use to convince ourselves, and hopefully the rest of the world, that our reasoning is sensible. This is all in service of what matters most to a social primate like us—not a philosopher's goal of metaphysical truth but sway.

Narrative is for making sense of the world. And it does so by simplifying the world. All narrative is reductionist. And once we have a narrative that's giving coherence and order to our existence, we typically defend it with half-blind vigor. To lose one's special narrative is like gravity suddenly switching off and meaning spinning away. This is a nauseating feeling, and most of us go through our lives making sure it never happens. We do so by devoting our mental resources not to testing our narratives but to protecting them.

It would be pretty to think that we at least *formed* our narratives rationally before getting so dogmatic about them. A rational process of narrative formation would go like this: we encounter facts as we move through the world and then we develop narratives to make sense of them. But the narratives in our heads are more like dingy paperbacks passing down the generations through used book stores. We pick up the book. We read the story. And it becomes our reality. Sometimes we may scribble little exclamations of dissent in the margins. Occasionally, some of us try to author whole new chapters in the fly leaves at the end. But most of us live primarily inside the stories we inherit.

When we say colloquially that "Sally has a narrative"—whether it's derived from Marxism, Islamism, feminism, libertarianism, or Flying Spaghetti Monsterism—we mean Sally has beliefs that are driven by a specific story of how the world got to be this way. And she also has narrative-based notions for how we should behave moving forward. But it's at least as true to say that *the narrative has her.* Once a powerful narrative colonizes Sally's mind, the narrative seizes the agency. She doesn't build the narrative out of facts so much as the narrative selects and shapes what Sally will accept as fact. I'm not denying that some narratives are far truer, and far more useful, than others. I'm saying we have an incurable tendency to cram more disorderly reality into our prefabricated narrative molds than we should.

During my last weeks of work on this book, for example, protesters who believed the 2020 presidential election was rigged fought their way into the US Capitol Building. A real-life, real-time Heider–Simmel experiment played out as tens of millions of Americans watched exactly the same chaotic footage and yet saw wildly different movies.

For people on the left, especially people of color, what first jumped out was the latest example of racist double standards in policing. The defenders of the Capitol, they claimed, treated the mostly white protesters with kid gloves, while they'd busted heads during the Black Lives Matter protests of the previous year. If the protesters had been mainly black and brown—according to news coverage, op-eds, and statements from politicians including Joe Biden—the police would have *obviously* unleashed hell. Why couldn't America see this?

To other people, especially white people and political moderates, this narrative of structural racism was a dud. That officers didn't violently repulse the crowd *obviously* wasn't a reflection of a racist double standard, just that a vastly outnumbered police force was being routed by many thousands of raging insurrectionists. First, the police get whaled on by rioters, and then they get whaled on again by pundits and the president-elect. Talk about insult to injury! Why did the Left have to make *everything* about racism?

Most Republicans watched the footage in horror as the protest turned into a violent riot. But many of them were equally horrified by their knowledge that Left media would—what's new?—lay all blame at the feet of their president. He told his supporters to protest a stolen election, not to riot! His comments were perfect! Just once, couldn't the media cover the president honestly?

A smaller group of Republicans saw no horror at all. They saw a stirring action film about the purest expression of patriotism since the Massachusetts militia rose against British tyranny at Lexington and Concord.

And finally, those in the flourishing QAnon wing of the Republican Party looked over the chaos with a knowing eye. Through minute study of the footage, they found irrefutable proof that the whole Capitol attack was a false-flag operation conducted by activists from Antifa and Black Lives Matter. The goal was not just to smear the president and his followers but also to diabolically frame patriotic *resistance* to a coup d'état as an actual *attempt* at a coup d'état.

My point isn't that the truth about this incident is unknowable. Whether it takes a year, a decade, or a generation, America's blood will hopefully cool, and we'll reach a consensus on this divisive episode, just as we have on others: OJ did it; the Iraq War was a bad idea; there was nothing good about the Capitol riot or the fictions that incited it.

My point is that the human mind abhors a story void. And so, when we see disorderly events unfolding in the world, we whip out our handy prefabricated narrative molds and wham them down over the disorder with terrific force. That, or we go to the information outlets we count on to wham our favored narratives down for us. Either way, we get stamped-out replicas of our favorite narratives, which mash down or cut away whatever doesn't fit. And in this fashion we use our narrative molds to *fabricate evidence for their own truth.*

I'm confident almost everyone can agree that this is pretty much how narrative psychology works. It's good we can agree. What's problematic is that almost everyone will also carve out space for one modest caveat: *the narrative I use to mold reality, unlike those others, is actually truly cast.*

Unfree Will

The Heider–Simmel effect, as it plays out in psychology labs or real-life situations like the Capitol riot, shows that we have little executive control over the stories that crowd into our minds when we look out on the world. This is disconcerting, and its practical effects can be terribly divisive. But something good can come out of knowing about the Heider–Simmel effect: a more constructive and compassionate way of seeing the narratives, and narrative-driven behaviors, we find objectionable.

Consider your own political orientation, which determines to a great extent the narratives that define your reality. Twins studies suggest that wherever you happen to find yourself on the Left–Right ideological spectrum, your genes are between 30 and 50 percent responsible.[4] And what explains the other 50 to 70 percent if not your environment—your culture, your family, and the stories you were raised on?

We are all products of some combination of genetics and social conditioning, which means that our personal tendencies and traits are not self-built. You didn't, for example, choose your brain. You didn't choose your parents or the genes they gave you or their approach to nurturing you. You didn't choose *not* to be born a psychopath, nor did you select *against* a horror-show childhood that might have turned you into one. You didn't choose *not* to be born into a situation of dirt poverty and ignorance that would have made it highly unlikely that you'd ever grow up to read this book. You didn't wisely select the genetically based personality traits—being warm or cold, stressed or loose, curious or closed off, impulsive or controlled—that made you whatever you are, from a lazy quitter to a gritty winner.

This isn't the place to slog through a debate about free will.* I'll just note that research suggests that our wills are far less free than most of us think. Here's the spooky truth: if you were hooked up to the proper laboratory gear, neuroscientists would be able to see your brain making a decision (say, to waggle your wrist) before you had any conscious awareness that the choice had been made. "One fact now seems indisputable," the neuroscientist Sam Harris writes. "Some moments before you are aware of what you will do next—a time in which you subjectively appear to have complete freedom to behave however you please—your brain has already determined what you will do. You then become conscious of this 'decision' and believe you are in the process of making it."[5] Again, this isn't the place to wade into all the complicated details, but these neuroscientific experiments line up with reams of research in the human sciences showing that our behavior is largely driven by subterranean impulses that the conscious mind has no awareness of.[6]

Most of us strongly resist the notion of unfree will, not least of all because it feels deeply at odds with the subjective feeling that we are in control of our own minds. People also worry that a doctrine of unfree will would be a slippery slope to moral relativism and even nihilism. There are indeed very heavy-duty moral implications to

---

* If you'd like to slog through these debates on your own time, I suggest beginning with two never-sloggy books: for the synoptic manifesto, Sam Harris's *Free Will*, and for the full eight-hundred-page scientific case, Robert Sapolsky's *Behave*. For a classic overview of complementary research on the unconscious drivers of human behavior, see Daniel Kahneman's *Thinking Fast and Slow*.

questioning free will. But they are all to the good. To quote Harris again: "I think that losing the sense of free will has only improved my ethics—by increasing my feelings of compassion and forgiveness, and diminishing my sense of entitlement to the fruits of my own good luck."[7]

Let me state my opinion as starkly as I can: All evidence suggests that we are no more responsible for the smart or dumb stories that fill our skulls than a depression in the earth is responsible for brimming up with pure or fetid waters. If this is true, it challenges some of the deepest and most precious illusions produced by narrative psychology. If free will (at least in the way we usually think about it) isn't real, then villains (at least in the way we usually think about them) aren't real either.

Look at it this way: We personally discovered almost none of the information that fills our heads. Our knowledge is overwhelmingly based on things that people have told us—much of it dubious. This means that people have "wrong" beliefs largely because they put their trust in the wrong authorities. We should therefore strive to disentangle our judgments of the truthfulness of a belief from our moral judgments of belief holders. It is natural for us to conclude that people who have bad beliefs are bad people. But this conclusion is a stark non sequitur. "Bad people," in the main, are simply those who had the misfortune first to encounter, and then to believe, the wrong storytellers.

This way of thinking strips us of simplifying and comforting fictions while depriving us of the ecstasy of sanctimony. But it leaves us with something better: a more merciful view of human action and with it a chance of getting through to each other. When "they"—whoever your "they" are—look out on the world and see stories obnoxiously at odds with your own, realize that they are perhaps to be pitied, sometimes to be feared, but not to be despised. And if you do, there's a better chance "they" will pay you the same courtesy.

The Storyverse

Leo Tolstoy celebrated art, especially his own form of story art, as "a means of union among men, joining them together in the same feelings and indispensable for the life and progress toward well-being of individuals and humanity."[8]

A century later, Mark Zuckerberg besotted himself with a similar dream. Facebook was infamously launched from Zuckerberg's dorm room as a way to rank the hotness or notness of female students at Harvard. But a few years later, this puerile and degrading technology had been retooled and rebranded as a bridge builder between family and friends. Soon Zuckerberg was talking about his site in frankly utopian terms. By connecting all of humanity in a single web and allowing us to share our stories, ideas, and feelings, Facebook would help us empathize with one another, old prejudices and misunderstandings would fall away, and the tide of harmony and happiness would rise accordingly.

In this, Zuckerberg was expressing a twenty-first-century version not only of Tolstoy's views but also of Marshall McLuhan's "global village."[9] McLuhan argued that the amazing new technologies of mass media—radio, film, TV, and print stories fired down electrical wires—were steeping the world's citizens in the same values and stories, drawing us all together into one human collective.

And for a time, at the height of the broadcast era in which tens of millions regularly watched the same TV dramas, listened to the same news and songs on the radio, and sat through the same films, McLuhan's vision seemed to be bearing out. Consumption of the same mass media drew us away from extremes and toward the middle.[10] For this, broadcast-era storytelling was widely condemned, especially by leftist intellectuals, as a technology of horrifying conformity. TV, in particular, it was said, zombified and brainwashed people into a white-bread conventionality. It's not that critics were wrong to be concerned; it's that they failed to see that the alternative could be worse.

Even at the height of the broadcast age, we didn't live in the same virtual village. But we did live in the same general *storyverse*, and it was hard to get out of it. "Storyverse" is my term for the mental, emotional, and imaginative space created by the stories we consume in any media whatsoever—from the bedtime tales of our childhood to the stories on Netflix, Instagram, and cable news, to the sermons we hear in our places of worship. Our storyverses aren't "reality" but our personalized versions of reality. The laws of these storyverses are often determined by a set of master narratives (or sometimes only one) that, like a decryption key, determines how we will make sense of everything around us.

Although few of us know it, most of us go through life all fouled up in the gluey webs of our storyverses—bending reality to our stories rather than the other way around. More troubling, humans can inhabit exactly the same physical reality but live in foreign storyverses. This may seem like a complicated idea, but anyone who's watched Fox News for a while, then watched MSNBC for a while longer knows what I mean: while we all think we live in the Right-Side Up, an awful lot of us are living somewhere in the Upside Down.

And you'll understand the storyverse concept even better if you've ever spoken to an acquaintance who's apparently *not* an unmedicated paranoid schizophrenic but is still engrossed in the mass hallucination of the QAnon conspiracy story, in which the forces of goodness are battling a satanic army of pedophile-cannibals bent on world domination. In either of these examples, we see communities isolated from one another in distinct storyverses, governed by alien physics, menaced by different villains, with different heroes racing to save the day.

By spreading connection and comity across borders, Zuckerberg's technology was supposed to signal the realization of McLuhan's global village. It has accomplished something like the opposite. It can't be absolutely proven because the data are merely correlational, but I don't think it's a coincidence that the rise of social media corresponds rather perfectly to a big bang of polarization and social instability, not just in America but in much of the world. Zuckerberg's technology, along with much else on the internet, ended up being a better enmity than unity machine—better at blowing up bridges than building them. As the computer scientist Jaron Lanier puts it, the rise of social media coincided with "an explosive amplification of negativity in human affairs."[11]

Here's one reason why: At any previous time in the last few hundred thousand years, if you could have floated over the Earth, you'd see people enjoying stories *together*—crowded around savannah campfires, theater stages, blaring radios, and glowing TVs. But if you could float over the roofs of modern homes (and had X-ray vision to boot), you'd see people consuming far more storytelling than ever before. But mostly they'd be doing it alone—staring at their screens or listening to podcasts through their earbuds. Story doesn't even bind single households together anymore, much less the whole globe.

In the so-called network era of television, when three broadcasters (ABC, CBS, NBC) dominated television production, shows had to be calibrated for diverse tastes. This created a market imperative for crappy "lowest common denominator" programs like *Alf* and *Touched by an Angel*. But it was cohesive crap.

Going from the pre-Guttenberg age in which formal storytelling was overwhelmingly consumed communally in smallish groups, to the mass audiences of the broadcast age, to the new age of story "narrowcasting" represents a sea change in human life. It amounts to a dangerous social experiment that seems to be going awry. Story has gone from being the great uniter, as James Poniewozik puts it, to the great divider.[12] Story used to drag us all to the middle and make us more alike. Now we're all in our own little storyverses, and instead of making us more alike, story makes us into more extreme versions of ourselves. It allows us to live in story worlds that reinforce our biases rather than challenge them. The end result is that everything consumed in our storylands just makes me more me, and you more you. It also makes "us" into more extreme versions of "us," and "them" into more extreme versions of "them."

The sharp balkanization of American liberals and conservatives—with all the dire consequences for civic harmony and national cohesion—is largely a result of each side's ability to live entirely inside the storyverses of the Left or the Right.

Endarkenment

Those who object to the notion that we're passing over into a post-truth world are apt to pound on the table, saying, "Can't you hear this table resound as I smack it? That's because my fist has truth. The table has truth! The physics that produces the sound wave is real! The biology that creates the pain in my bones is real!"

Yes, reality is real. Truth is true. But what does it matter if no one can agree?

Doubt is wholesome and good. Doubt makes us prudent. Doubt makes us humble about our claims and charitable to alternative points of view. Doubt is a prophylactic against fanaticism. A doubting world is a better world.

But as it turns out, the post-truth world isn't one where most people stop believing that the truth is out there—where they become beret-twirling, Gauloises-puffing postmodern relativists. On the contrary, the post-truth world is a world of greater certainty. It's a world where, no matter which bonkers story you believe, you can back it up with loads of information that resembles actual evidence.

A post-truth world is a world where evidence is stripped of power. Moving deeper into post-truth territory is scary because it was a commitment to evidence, above all, that freed us from the Dark

Ages by weakening the dominion of story. There was an optimistic sense in the Enlightenment that the world ran on logical principles and that even though the truth was shy and evasive, communities of principled people using powerful techniques and methods could corner it and make it speak. This was the invention of real science. And this pushed the darkness back. It emancipated us from the tyranny of stories (most particularly religious fundamentalism) and brought on the expansion of prosperity and the reduction in suffering that we've known since the Enlightenment.[13]

It's evidence that allowed us to crawl out of the Dark Ages and into the light. It's science. Now we're leaving behind the world of shared reality. And we're entering a dreamland where truth is decided based on what's the best story—or which story is backed by the most muscle—not on what's backed by the best evidence.

This is a scary prospect. It's the candle of the rational Enlightenment beginning to gutter out and a foretaste of a new endarkenment bringing renewed fervor to our prejudices, our superstitions, and our bent for tribalistic violence. The danger is helpfully personified in the great hero of this age of fiction and, thus, inevitably, of this book. He's the living incarnation of all the power of our story psychology and all the danger. Perhaps you've sensed him all along, standing just offstage, panting with frustration at being held back for so long behind a curtain of boring ideas. He's panting with desire to bask at the center even of this little stage.

It's time now. Cue the Big Blare.

The First Fictional President of the United States

For all but the last two weeks I worked on this book, my favorite fictional character lived at 1600 Pennsylvania Avenue in Washington, DC. I call him the Big Blare because, well, he's big, of course. Big in voice. Big in height. Big in circumference. And because he blares. I call him the Big Blare because he draws attention in the same way that a blaring horn, or a splash of light in a painting, hijacks the central nervous system and forces it to pay attention. The blaring fluorescence of his skin, and hair, and voice, and views are impossible to ignore even when we want to.

The Big Blare is a lustmonster famished for "hamberders" and gold and beauties to display on his arm, and above all to have the whole world share in his self-regard. Or if we can't give him the love he deserves, we at least owe him the respect of our hate. Anything but the indignity of having a single Earthling go a whole day, a whole hour, a whole minute without seeing his face or speaking his name.

And this is why his is the name that shall not be written by me. The whole reason he became president, and the whole reason he gets out of bed every morning, is to blare at the world—to demand that we look at him and say his name. It's the same motive of the arch troll and the school shooter.

The Big Blare was underestimated as a presidential candidate largely because pundits couldn't *detect* his greatness as a popular storyteller—in the same way literary critics can't *detect* the talent of a much-scorned popular novelist like James Patterson. All pundits could see was the shocking ineptness of his style of presentation. But it was an ineptitude that somehow riveted everyone's attention, whether they were gazing on in ardor or paralytic horror. And despite all the rambling and backtracking he still delivered a clear, simple story that seethes with primitive emotion.

The Big Blare's story, "Make America Great Again," has mythic sweep. Through cowardice, corruption, and conspiracy, the world's greatest civilization had fallen to carnage—multihued barbarians were spewing out of the world's shitholes, crashing through the nation's gates to corrode it from the inside. But America was poised to rise again. All we needed was a culture hero to lead us back to the age of gold.

The Big Blare happens to be the type of fictional character I love best, the comic grotesque: the great Ignatius Riley from *A Confederacy of Dunces*; the rotten protagonist of Martin Amis's novel *Money* or his dad Kingsley's novel *One Fat Englishman*; the type you might come across in a Tom Wolfe novel or a farce starring Will Ferrell or Danny McBride. The Big Blare is, like these characters, an implausible combination of greed, vanity, depravity, solipsism, selfishness, and insecurity all overlaid with an out-of-control case of Dunning-Kruger–style narcissism. And this perfect storm of psychological pathologies is wrapped up in a physical package so viciously exaggerated that it violates the rule that fictional characters should have some type of real-world plausibility.

My only problem with the Big Blare is that he somehow escaped the fiction world he was clearly made for—an inspired satire of the crudest elements of the American character all personified in one tumescent, ochre-tinged symbol. And when he escaped, he brought the fiction world with him and marooned almost half of the country inside it.

Of course, presidents are *depicted* in fiction all the time, down a spectrum from idealized heroes to satirical buffoons. And it's true that every presidential candidate creates a persona and shapes a truthy narrative about his or her life and vision. But there are, as the saying goes, levels to this game. That we actually elected such an unabashedly fictionalized human as our president is the ultimate symbol of our passage out of the "reality-based" world into a "post-truth" world. Yes, the Big Blare has an undeniable physical reality. But from the very beginning of his life on the public stage, he's been curating a fictional version of himself as a self-made tycoon and vigorously thrusting that fiction—in his books, his reality shows, and the 30,573 fictional claims he made during his presidency[14]—at a half-willing world.

The Big Blare is a postmodern character who's furiously improvising his own nonstop reality show even as he stars in it. And he's without question one of the most successful storytellers, and influential characters, in world history. Over the years since he first announced his candidacy, no human being has ever had their face seen more widely or their name spoken more frequently. According to a statistical analysis of web traffic and social media, he became the most famous man in the world sometime in 2016.[15] And he's much more famous now.

The Natural

The Big Blare show, running nonstop since the announcement of his candidacy in 2015, is a massive and still ongoing Heider–Simmel experiment where people watch the same footage and see films with crisscrossed heroes and villains and diametrically opposed morals of the story. It's as if one group of people were convinced they were watching a terrifyingly dystopian version of *Anchorman* with Will Ferrell, and another group of people watched the same footage and saw not scary buffoonery and thuggery but hope and majesty.

Where some see history's most flagrant and talented serial liar, others see the one man brave enough to say truth. Where some see a vandal trashing cherished norms, others see a rebel willing to break things to fix things. Where some see the gravest threat to American democracy, others see democracy's last hope.

But it's important to recall that the Big Blare wasn't always such a polarizing figure. Virtually everyone across the political spectrum initially saw the Big Blare's presidential run not just as a longshot but as a joke. Then the Big Blare took the whole conflict model of storytelling and applied it to his presidential run. Like a contestant on a reality show, he realized that the most reliable way to win screen time is to be the heel—the bad guy who wreaks havoc and drives conflict. He gambled that a strategy of *conflict forever, comity never* would be irresistible to the news media.

He was right. Early in the Big Blare's run, news organizations turned away from the other personalities in the Republican primary—including many experienced and responsible politicians—because they were unforgivably boring. Every network and newspaper made the Big Blare the lead character in an unfolding political drama, and he obliged by committing new outrages, driving new storylines, nearly every hour of every day. This was no slow-drip literary novel. It was an over-the-top political soap opera with a wildly twisting plot. The Big Blare's presidency was a result of his ability to hack narrative psychology and get media organizations, composed overwhelmingly of people who despised him, to gift him many billions of dollars in free advertising.

I can't quite blame the news media for making the rise of the Big Blare possible. They couldn't help themselves. Journalists are storytelling animals extraordinaire, and he's an irresistible character churning out endless fodder for big, rich stories. Moreover, we're all just as complicit as the news media. Fast-food companies don't stuff us full of fat and sugar against our will. They feed us what we like. And there's a reason why journalists stuff us full of the fatty, meringue-topped confection that is the Big Blare. We all love the Big Blare show, even if we dislike him. And the numbers prove it: The Big Blare and the media did outstanding business together. Both parties got what they wanted. The Big Blare became the universe's largest black hole for sucking in mass quantities of human attention. And struggling media outlets got a plot machine that earned them a lot more traffic and a lot more money.[16] And in the end they made a man of Kardashian-level gravitas and cultural significance into the most important and dangerous man on Earth.

When he's gone there will be a Big Blaring absence in your life. The more he disgusts you, the more you'll miss the meaning he gave you. You'll agree with me, if only guiltily, that he's your favorite fiction character too. No one more outlandish. Everything so *much*—the hair, the tan, the voice, the petulance, the vanity.

You'll miss your villain. You'll feel glum and flat and you won't even know it's because you're jonesing for the rage spike you got every time he showed up in your news feed. And after he's gone, you'll be constantly trying to get back in the simple story he gave you about good guys like you and your friends up against the bad guys in red hats. And you'll try to get back in that story by consuming actual satirical fictions about his presidency and even crazier histories and biographies.

The Big Blare is the great exemplar of the aphorism "the storyteller rules the world." Because it was his instinctive gifts as a fantasist and his unstable genius in the control of narrative that allowed him to gain the closest thing to rulership of the world that currently exists: the presidency of the United States. But the Big Blare is the farthest thing from the type of storyteller-king Plato dreamed of. What Plato wanted most was someone like the current president of China: a person who's been cultivated throughout life by a council of elders to rule with maximum rationality. What Plato feared most was the type of storyteller-king the Big Blare represents: a born demagogue who has no philosophy beyond self-interest and is entirely governed by emotion and appetite.

The Big Blare's rise as storyteller-king is far more impressive than that of any Soviet premier or the Kim Dynasty in North Korea or Putin or any pope. This is because the Big Blare is constrained, in the building and maintenance of his personality cult, by the Constitution's guarantees of a separation of powers and a free press. Unlike previous storyteller kings, he had nothing approaching a monopoly on the channels of storytelling. His stories had to actually *work*, whereas in most kingdoms of story, citizens duck the noose by convincingly *pretending* the stories work even when they don't.

The Big Blare had to actually compete and win in a totally free market of storytelling. And here's the amazing part: he did. Especially at the start of his 2016 run, it was him against the whole ganged-up world of respectable storytelling—liberals *and* conservatives—and he trounced us.

P.S.

It hasn't escaped my notice (though honestly it almost did) that I've violated my own principles not only by casting the forty-fifth president as a villain but also by doing so with echoes of his own flagrant and sophomoric cruelty. A research-intensive book of this sort is written over a long period of time as an author learns and his or her thinking evolves and hopefully matures. At the end, a writer will usually double back over older pages and bring it all up to the standard of the newer. In so doing, the writer produces the illusion that the book leapt from his or her head all at once, fully formed, rather than being a long slog toward clarity that took months or years.

What I wrote about the forty-fifth president was roughed out fairly early in my process and then updated as events unfolded. Much of what I wrote about the desirability of producing history without villains was formulated nearer the end. In my last revising passes, I softened my portrait of the Big Blare. But I opted not to fuzz out all the harsh lines in order to illustrate a point I'll get around to making more fully in the conclusion: It's hard to jar narrative psychology out of its deepest, oldest grooves. It's easy to preach reform in how we construct narratives and hard to move from preaching to practice.

I finished composing my portrait of the Big Blare prior to the 2020 presidential election. Since then, he's continued to churn out fodder for new stories with shocking new twists. He spent the interregnum before Joe Biden's inauguration improvising his way through a massively seditious conspiracy story that went something like this: *the election was beyond doubt stolen from me but by enemies so cunning that they left no trace of evidence.*

Despite Republican and Democratic judges flinging out dozens of the president's legal challenges, his story has successfully suspended the disbelief of 70 percent of Republicans, who express skepticism about the election's integrity to pollsters. As we've seen, some believed this story so fervently that they mobbed, sacked, and pillaged the US Capitol Building.[17] And as I'm finishing this book in late January 2021, twenty-five thousand National Guardsmen are stationed around the Capitol in an effort to deter an insurgent army of fellow Americans from attacking the government. The worries I entertained in the previous chapter about a dangerously escalating civil war of stories no longer seem hysterical.

That the Big Blare lost is a good sign for the reality-based world, but only up to a point. Just think of what it took to beat him. It took the largest voter turnout in American history to squeak out a narrow Electoral College win (the contest would have ended in a tie if Biden had won just forty-six thousand fewer votes across the swing states of Arizona, Wisconsin, and Georgia). And this was in turn driven by an uprising of reality, in the form of Covid-19, against the Big Blare's fictions. Without the gross incompetence of the president's response to a one-hundred-year plague, it's likely that he would have won a second term.

And even now that he's left the White House, the Big Blare will continue to rule red America as the storyteller-king in exile, exerting enormous sway over our lives and fortunes for as long as he (or his movement) lasts. Historians may remember him less for specific actions in office, not even inciting the insurrection, than for teaching American politicians how frail and weak reality is and how easily it can be overpowered by the right type of fictions told with all-in commitment. Moreover, he believes he can win the presidency again in 2024 for the same reason he won in 2016: journalists won't be able to resist making him the star of their dramas.[18]

Even if a second term seems less likely after the insurrection, a second impeachment, and the loss of his Twitter megaphone, the Big Blare still has a 75 percent approval rating among Republicans. This makes him if not the once and future storyteller-king of red America, then the kingmaker. The way it looks now, either the Big Blare will be the 2024 Republican nominee or he will choose the person best suited to rule in his image.

An Academic Reformation

Plato was wrong about many things, but there are reasons why his frustrating, confusing, and maybe-just-somewhat-wicked book has been so revered. *The Republic* shoots for something we all want: a world logically engineered to achieve the greatest good for the greatest number. We all love stories, but almost no one actually wants the best storytellers to rule the world by ginning up our emotions and depressing our logic functions.

We need more reason in the world. But the main way we get it won't be through an assault on story. Plato wanted to weaken the dominion of storytellers by banishing any tale-teller who would not turn his quill to the needs of the state; he wanted to burn or bowdlerize everything written in the age of free poets; and he wanted strong state control over the style in which stories were told, with prohibitions on the whole bag of emotion-evoking tricks tale-tellers use to make a story come alive in our minds. Variations on this experiment have since been tried out in all of history's finest dystopias—North Korea, the USSR, Maoist China, Nazi Germany, Cambodia under the Khmer Rouge—and it has always ended in epic tragedy.

To establish a civilization where people live in the same, reason-based storyverse, the move shouldn't be to weaken our connection to story but to strengthen the shaky counterbalance to story, which is the logistikon. Above all, we need to double down on our commitment to science because science is *for* standing up to stories.

Science is the one republic that followed Plato's first impulse and tried to banish the tale-tellers. Storytelling was okay for teaching nonscientists, but it was rigorously suppressed in the journals where scientists do the slow, fussy work of trying to separate the true from the false. Narratives appear in scientific papers, but mainly for the purpose of exposing them to potentially falsifying tests. What science is, at bottom, is the most reliable method humans have devised for finding out which of our narratives about reality are true and which are false. Of course, no one thinks that science, an institution run by flawed humans, is perfect. But it's also true that even the harshest critics of science won't claim that they'd prefer to return to a prescientific age. Science is the one tool that forces us to see what's actually in front of us, rather than what our egos and stories would like us to see. Science is the most powerful way we have of keeping stories from rioting without restraint in our lives.

And the extent to which we're able to survive this post-truth moment and return en masse to the real world will depend on whether we can move back to a world where science, and other powerful forms of empiricism, regain their authority. And for this to happen, our primary truth-telling institutions have to change.

Academia and journalism play roles that are truly indispensable to democracy. Journalism tells our true stories about what's happening day-by-day in the world, and the university system is filled with enormously educated scientists and scholars tasked with giving us our most rigorous and trustworthy assessments of the truth about economic systems, gender arrangements, the contours of human psychology, the variables behind climate change, the function of art, and so on.

If journalism and academia are operating as they should, they can act as arbiters in a democracy's story wars. Without institutions that can provide trusted information to settle disputes, there's no way to end a story war short of real fighting or a cold war that freezes society in place and leaves it helpless in the face of its problems. This truth-seeking mission is the sacred and indispensable social role of academia and journalism. And our steep descent into a post-truth condition couldn't have happened if we were doing it well.

Just because the notion of liberal bias in mainstream media is a right-wing political slogan doesn't mean it's untrue. Registered Democrats (28 percent) working in the mainstream media outnumber registered Republicans (7 percent) by a factor of four,[19] but few doubt that the great majority of unaffiliated journalists actually lean left but prefer not to advertise their politics. I couldn't find statistics, for example, on the political affiliations of the workers at National Public Radio, but if you listen to stories on NPR as I have done every day for decades, you'll reach the conclusion that for all the diversity they seek to promote, it's almost inconceivable that a single actual Republican works in the content-shaping positions of reporting and producing their news programs.

But I want to focus on the failings of my own tribe, the academics, where the problem of ideological bias is even more extreme. In the 1960s, for example, American history departments already slanted decisively left, with 2.7 Democratic historians for every Republican. But in a recent study of almost eight thousand faculty at forty leading American institutions of higher learning, the ratio of liberal to conservative historians had risen to an incredible 33.5:1.[20]

According to research by the legal scholar Cass Sunstein, if you get a random selection of people in a room and have them argue a divisive subject, they generally work their way toward some sort of compromise position. Partisan beliefs, attitudes, and actions are moderated by the reasoning of the other side. But if you get a bunch of same-side people in a room and have them discuss divisive topics like abortion or gun control, they don't gravitate toward a median position. When homogenous groups are insulated from skepticism and counterarguing, they stampede toward the most extreme position in the room. When you have a room full of partisans, the question asked is seldom "Are we going too far?" It's usually "Are we going far enough?" This tendency is so strong and predictable that Sunstein calls it "the law of group polarization."[21]

History departments are extreme. But the demographics are tilted in every academic field. In a country where more Americans self-identify as conservative (about 37 percent) than liberal (about 24 percent),[22] studies suggest that registered Democrats outnumber registered Republicans on college faculties by somewhere between 8 and 13 to 1, with eye-popping imbalances in anthropology (42:1), English (27:1), and sociology (27:1).[23] In the field with the most viewpoint diversity, economics, conservatives are still outnumbered by liberals 4.5:1. In some fields, the ideological outlook is so rigidly uniform that it's difficult to imagine that a single "out" Republican could exist in the entire discipline. And, in fact, a study of fifty-one of the top sixty liberal arts institutions in America showed that there was not a single registered Republican in the fields of gender studies, peace studies, or Africana studies.[24]

Academics have been aware of these imbalances for decades; they just don't care about them very much. When I raise my concerns with colleagues, the most common response is a shrug and an only-half-joking "Well, how can we help it if the truth has a liberal bias?" But if we're actually interested in getting nearer to the unbiased truth, these imbalances are a catastrophe that undermines all the research and teaching we do. To be clear, I'm not saying this is a problem of bad optics (though that too). I'm saying that academic research appears to be biased because it most definitely is—systematically and in one direction.

The ideological homogeneity of academia is compounded by an increasingly ruthless authoritarian streak with sacrosanct, unchallengeable dogmas especially surrounding identity issues like gender, race, sexuality, and so on. An intellectual atmosphere of terrorized conformity—with all the mobbing, deplatforming, canceling, and proscription of forbidden questions—contributes to a post-truth world as much as Big Blarism does.

I didn't set out in this book with plans of writing something feisty or polemical. But I've still managed to tread on powerful toes.

I've felt no compunction, much less fear, about offending the sensitivities of the Catholic Church, religious people generally, conspiracy theorists, the tech oligarchs of Silicon Valley, the Big Blare, or the tens of millions of armed-to-the-teeth Americans who still love him. I even passed some pleasant hours drawing a satirical portrait of ISIS fighters LARPing their way through end-times make-believe (it ended up on the cutting room floor because its point was redundant, not because I was scared).

What has troubled my sleep is the prospect of saying anything even mildly heretical regarding the Left's sacred narratives. Like all contemporary writers, I'm aware that I'm always one loose move away from sparking my own auto-da-fé, knowing that it will be my liberal friends—not fanatic right-wingers—who will come running to the fire, shaking their cans of gas. This puts a certain English on virtually everything you read and governs what writers dare say and how we dare say it. No totalitarian regime can burn all its heretics, but conveniently all they have to do is burn a small proportion and the remainder will fall in line. When sufficiently terrorized, thought criminals burn their books proactively—or at least the ideologically incorrect chunks of them. This is another way of saying that academia claims absolute fealty to the noble ideal of freedom of thought and expression while still neatly deterring unorthodoxy.

To proceed as though intolerance, intimidation, groupthink, and homogeneity of ideology would *not* lead to timid party-line scholarship (a sort of miniature version of Soviet Lysenkoism) would be to yield to the epistemological equivalent of flat earthism. It would be to cluelessly insist that the facts academics have ourselves uncovered about the deep biases of human cognition and ideology don't apply to us.

But of course the public knows they apply to us and they discount their confidence in our work accordingly. Thus the academy has ruined its power across huge swaths of this country, with most Republicans (59 percent) now saying that higher education is a force for ill, not good, in America.[25] This means that any research on a potentially partisan subject—which means most research—will be treated as suspect regardless of its quality. Owing to overpowering ideological bias in the professoriate, the public will have excellent cause to doubt the gender studies professor talking about the causes of the wage gap, to doubt that a sociologist has asked all the pertinent questions in a study of police brutality, and to doubt the underlying motives of historians offering expert testimony at an impeachment hearing.

And they will also have reason to doubt science. The strength of science resides not in individual geniuses like Newton or Darwin but in an ingenious collective process, which—with its demands for mathematical rigor, peer review, and replication—guards against predictable human failings from outright lying to unconscious prejudice. Though the scientific method is a prophylactic against bias, it's not a perfect one. Academic scientists, too, lean strongly to the left, and we must assume that this shapes the questions they ask and the interpretations they favor. The consequences of this are presumably less distorting the farther you get from human-related fields (e.g., astrophysics) and more distorting as research presses up on areas of live societal concern (e.g., genetics, sociology).

Even in the event that scientists could learn to fully isolate their politics from their research, science's partisan tilt would still have disastrous PR consequences. For instance, the scourge of global warming conspiracy tales, which have all but negated our capacity to deal with the existential threat of climate change, are fictions authored by right-wing propagandists. But the fictions caught on partly because a verifiably biased liberal academy (including an overwhelmingly left-leaning community of climate scientists) made the conspiracies *seem* plausible.

Democalypse

I'm smiling idiotically. I'm watching a video showing a sequence of faces smiling back at me as they morph from male to female, brown to white and back to a different shade of brown. I watch a young woman's perfect skin go dry as she ages into an old man. Along the way her hair burgeons beautifully and then recedes to almost nothing. Personalities are carved into smile lines and expressed through subtleties of jewelry, clothing, makeup, and hair style.

I'm smiling idiotically because it's a heart-melting portrait of our species, showing the paltriness of our differences compared to our shared humanity. But my smile is also idiotic in a more literal sense because I know the faces onscreen aren't real—just ultrarealistic golems conjured by powerful computers. I know they're fake only thanks to the headline of the *New York Times* article I'm reading: "Do These AI-Created Fake People Look Real to You?"[26] For now, trained experts can still spot the miniscule AI tells, like minor anomalies when depicting jewelry, eyeglasses, and backgrounds. But every moment, the face-building technology is learning and improving.

Plato's *Republic* references an "ancient quarrel between philosophy and poetry."[27] The conflict is based on the storytellers' desire to spin out unfettered narratives and the philosophers' desire to build walls of truth to contain them. If the balance between the truth-telling philosophers and the fabricating storytellers is broken in the latter's favor, Plato warns, emotion overpowers reason and society loses its wits.

New technology, especially bullshit-super-spreading social media networks, is tipping the balance of power against the truth-tellers. And now we have the dawn of so-called synthetic media, often referred to colloquially as *deepfakes*, which presage a world in which any text, audio, or video evidence can be convincingly fabricated.[28] I worry that we may look back at the dawn of deepfakes as a crucial hinge point in human history: the moment reality died because technology severed the ancient linkage between seeing/hearing and believing.

Any antiquarian who stumbles over this book a generation or two hence may find the concerns about political bias in our truth-finding institutions, outlined in the previous section, to be cute (*Aw, humans used to worry about that?*). Technologists warn that we're racing toward an infocalypse—an information apocalypse. Because it will be possible to fabricate evidence of *everything*, it will also be possible to dismiss evidence of *anything*. The evidentiary foundation for rational decision-making will be blasted apart.

The infocalypse would make every other sort of apocalypse immensely more likely. If we can't agree on what constitutes real danger, how will we rally to confront it? This epistemological crisis may therefore be the greatest existential crisis we have ever faced, because it subsumes most of the others. Our feckless response to climate change, driven by the fact that so many of us found so many ways to disbelieve the experts, may only be a foretaste of what's to come.

To focus on just one matter of concern, observers worry that the infocalypse could also spell a democalypse—the end times for liberal democracy.[29] Writing in the giddy aftermath of the fall of the USSR, the political scientist Francis Fukuyama noted that liberal democracy was thriving across the world while various types of autocratic systems—whether monarchical, communistic, or fascistic—were rapidly going extinct. Fukuyama famously asserted that we were living at "the end of history."

Fukuyama wasn't claiming that natural disasters, wars, economic cataclysms, and cultural upheavals would stop happening. Rather, Fukuyama believed that political systems thrived or failed according to the same principle as biological evolution: survival of the fittest. For political scientists to worry about democracy being outcompeted by autocracy would be almost as pointless as worrying that a lower form of primate—baboons, for example—might one day leapfrog over humans.

Fukuyama's thesis hasn't aged well. But it gained so many adherents not because it was brash but because it was conventional. Nearly everyone in Fukuyama's audience already believed that democracy was superior to its competitors in both moral and practical terms. As Churchill famously quipped: "No one pretends that democracy is perfect or all-wise. . . . Democracy is the worst form of government except for all those other forms that have been tried from time to time."[30]

In support of Fukuyama and Churchill, if one thing marks hunter-gatherer politics, it is an adamant hostility to authoritarian leadership and an insistence on consensus-based decision-making. So, there is indeed a sense in which a yearning for freedom, and a hatred of domination by big men or cliques, is natural to human beings. It wouldn't be far wrong to say that humans in "the state of nature" are a democratic animal.[31]

But the state of nature (what evolutionary biologists call "the environment") isn't eternal or stable. And when the environment changes, organisms (or political systems) must adapt or die. Just such a change occurred beginning roughly twelve thousand years ago with the Neolithic Revolution. The advent of agriculture allowed humans to live in much larger groups, and the egalitarian systems of hunter-gatherers have been in retreat ever since.[32] Since the agriculture-driven population boom, various kinds of authoritarians have dominated human existence: chieftains, feudal lords, kings, popes, caliphs, and emperors.

For a brief moment in Greece, long-dormant democratic impulses flared up, then all but disappeared from Western society for two thousand years. It wasn't until the Enlightenment that a new form of democracy emerged and gradually proliferated around the world. When seen over the long view of human history, then, it would be wrong to say that democracy is an eternal system or that it will always outcompete other systems. It would be truer to say that, when it comes to governing large human collectives, various forms of authoritarianism have been much more prevalent and stable over the long run. And if democracy survives into the future, it won't be because it's inherently superior but because we who love democracy recognize its delicate contingency.

Modern democracies hew to the principle that the antidote to bad speech (like hate and lies) is better speech (like love and truth). In America, in particular, economic markets may be regulated, but the First Amendment guarantees a radically libertarian marketplace of ideas and narratives. This reflects our faith that when it comes to the marketplace of narratives, the good and true eventually beat out the bad and false—just as we believe good widgets beat out bad ones in a free economic market.

But mounting evidence suggests that, in a post-truth marketplace of ideas, misinformation and disinformation overpower truth with worrying predictability. In a disturbing article in *Science*, a team led by Soroush Vosoughi shows that false narratives don't just beat true narratives but utterly trounce them on every metric of spread on social media platforms.[33] As a timely example of this, a different team of researchers studying information on Covid-19 found that "content from the top ten websites spreading health misinformation had almost four times as many estimated views on Facebook as equivalent content from the websites of ten leading health institutions, such as the World Health Organization (WHO) and the Centers for Disease Control and Prevention (CDC)."[34] The confusion sown by these misinformation sources hamstrung attempts to minimize the pandemic's human and economic casualties. The virus spread so well in America, in other words, precisely because fiction outspread the truth.

William Butler Yeats's poem "Things Fall Apart" conjures a blood-soaked vision of a world spinning harder and faster until the center cannot hold and "mere anarchy is loosed on the world." The nonstop, never-off division and misinformation machine of social media is, more than any other thing, the centrifuge in which we twirl.

Like the robber barons of yore, the leaders of companies like Facebook and Twitter have created enormous value but have also imposed devastating externalities on all of civilization. In the age of robber barons, the negative externalities included exploitation of the workforce and environmental degradation. In the age of social media, the platforms have mainlined virulent social carcinogens—hatred, division, delusion—straight into the bloodstream of the body politic.

Reformers have a solution for this: regulate social media and consider breaking up the biggest platforms.[35] Yes, please, let's try. Social media is scarcely more than ten years old, and it's rushed over us like a hurricane. It's far too soon to conclude that we can't get it under control and rebuild what's been wrecked.

But there's reason for pessimism, too. Facebook, for example, is in large part a narrative-distribution platform. In fact, Facebook is easily the largest and most powerful publisher in history, distributing content to three billion humans and rising. It hasn't thrived because it discovered a new way of capturing attention. In large part, its algorithm just independently discovered the oldest way of capturing attention—the universal grammar of storytelling—and figured out how to distribute it on a colossal scale. The intelligence behind the algorithm may be artificial, but the narrative psychology it exploits is entirely natural.

To wish the negative externalities of Facebook away is a near thing to wishing away the universal grammar of storytelling. It's to fantasize that social media companies are *creating* demand for dark, divisive, and morally provocative material rather than responding to it. It's therefore to fantasize that a different algorithm serving as a router for narratives of truth, goodness, and positivity could perform almost as well. But no matter the business model (free, subscription, or whatever), social media platforms will naturally conform to the built-in regularities of narrative psychology, whereby the darker the narrative, the more it crackles with moralistic energy, the more likely it will win out in the story wars.

Plato's Republic of China

In a paper on how story shapes group identity, the psychologist Lucas Bietti and his colleagues argue that "storytelling is arguably the primary social activity by which collective sensemaking is accomplished."[36] In other words it's how a whole society gets on the same page about everything that matters.

Twenty-four hundred years after Plato wrote *The Republic*, the technological conditions finally exist for realizing his dream. With increasingly infrequent recourse to hard power, the storytelling kings of the Chinese Communist Party will be able to sculpt a national storyverse to gain compliance from its citizens. This project of total state surveillance, tied to total control over all channels of media, has already reached a fairly mature stage. As the producers of a recent episode of *Frontline* say, "A new form of governance being developed [by the Chinese] to control humans through technology, which is already being exported around the world, enables authoritarian governments to control their citizens to a terrifying degree. The central ideological battle of this century will be China's model of authoritarianism versus the West's increasingly shaky liberal democracy."[37]

And who would bet on us coming out on top?

"Great nations require national unity," writes the China scholar Liu Mingfu.[38] The contest among nations is, in large part, a unity contest. And when it comes to national unity, the Chinese have major built-in advantages. They have ethnic unity, with 95 percent identifying as Han Chinese. They have a five-thousand-year history as a civilization. They have a shared mythology given to them by the Communist Party. And, in contrast to the fractious individualism of Western (and especially American) culture, they have a more collectivist cultural orientation that prioritizes the group over the individual.[39] On top of these built-in unity advantages over the Western democracies, the digital tools that are so threatening to us, and that practically guarantee a future scorched by increasingly intense and destructive internal story wars, practically guarantee story peace for the Chinese.

The tools of story war are already being weaponized by the Chinese (and other authoritarians) against the West, while their own carefully constructed storyverse is protected behind the rampart of the Great Firewall. Deepfakes aren't a threat to the Chinese. On the contrary, deepfakes look more like the realization of the platonic/totalitarian goal of suspending a population inside any fictional dream that rulers can imagine. One should perhaps try to be open-minded about all of this. It's possible that the matrix the Chinese build *won't* be a science-fiction nightmare. After all, even in *The Matrix* film, the simulation was a pretty good place to live. What's really awful in *The Matrix* isn't the simulation but being flushed into the hell of unvarnished reality. Moreover, whatever the Chinese build may not be as unpleasant as what seems to be in store for free societies, which isn't people living in one matrix, as in China, but people living in a variety of mutually incompatible matrices locked in wars of stories escalating toward something worse.

CONCLUSION

A Call to Adventure

ROBERT PENN WARREN (1905–1989) IS BEST REMEMBERED AS the Pulitzer Prize–winning author of *All the King's Men* (1946). But he was even more accomplished in verse than prose, claiming two more Pulitzers for poetry and being honored as the first poet laureate of the United States. A snip from his great narrative poem "Audubon" (1969) serves as the epigraph to this book:

> *Tell me a story.*
> *In this century, and moment, of mania,*
> *Tell me a story.*

Warren was writing from within a moment of mania within a century of the same. He wrote his poem—about the great artist and naturalist Jean Jacques Audubon (1785–1851)—at the maniac peak of the Vietnam War. It was a time of American division and cultural upheaval unparalleled since the Civil War. And that moment of mania was embedded in the even larger manias of the twentieth century. Most of the audience for Warren's poem would have lived through WWII, and many, like Warren himself, would have experienced World War I and the Depression, too. And all his readers would have been living through some of the iciest moments of the Cold War, when every day that didn't end in nuclear Armageddon was a lucky one.

Warren closes the long poem by begging to be told a story of "deep delight" that will perform the most profound alchemy of narrative: transmogrifying chaos into order, mania into meaning.

Of course, people seem always to feel that they're living in ages of mania and maniacs, with the end times closing in. And they have always begged for stories to make the madness meaningful and consoling.

But our moment of mania really does seem special, if only because we can't turn to our storytellers to save us. Thanks to the speed of cultural and technological change, the stories that once helped to lift us up and join us together are causing the mania in the first place. The cure to dissolution and disorder has become the cause of it.

So, here's the question this whole book has been building toward. As story, and storytellers, grow stronger and stronger, and as the power of facts and evidence grows weaker and weaker, what can we do to save ourselves from societal schizophrenia, in both dictionary senses of being split against ourselves and living inside dangerous collective delusions?

Only one thing's sure: it won't be easy.

Allegory of a Cave

About fifteen thousand years ago in the foothills of the Pyrenees, in the land we now call France, a man stood knee-deep in a small river, staring into the wet mouth of a cave.[1] The river flowed from the cave, carrying the muddy scent of the earth. The man waded forward into the deepening water. As he entered the blackness of the cave, he leaned into the current, holding his sputtering torch up high while the river soaked his beard. After a hundred feet, he clambered onto a thin gravel beach, where he stood for a time, teeth chattering in the bubble of torchlight.

Holding tight to the wall and crunching the gravel with his feet, he moved deeper into the cave. He looked up to see stone daggers hanging from the ceiling and the shapes of ibexes and bison gouged into the walls. He reached a vertical shaft that stretched upward for forty feet. It was full of branches that men had forced cross-wise into the tunnel, and he climbed the branches like the rungs of a ladder.

He emerged in a space so cramped that he couldn't stand. So, he pressed forward on his hands and knees, the walls tightening around him. At times he could rise up on his feet. At times he had to slither on his belly, angling to slide his wide shoulders deeper and deeper into the blackness. He skirted pitfalls that opened up at his feet like deep, hungry throats. He passed the bones of cave bears. He passed into little rooms where the quartz rock shone like stars.

He reached his destination: a chamber shaped like a huge bowl tipped upside down. The man took hollowed rocks from his bag. The hollows were white with congealed fat. He lit the wicks that were standing in the fat, and the lamps sputtered and smoked and cast out their light. The man then kneeled at one end of the room and used rocks to dig and scrape at the clay floor. He pried up thick slabs of clay and carried them to the other side of the room. He leaned the slabs against a low boulder and then he began to work the clay with hands and stones and the horn of an ibex, polishing it all smooth with puddle water. He shaped two clay bison, a bull rising up to mount a cow. The sculpture stood unseen in the blackness for many thousands of years until it was rediscovered by adventuresome boys in 1908.

The amazing thing about this little story is that it's mostly true. About fifteen thousand years ago in France, a person or persons really did swim and climb and crawl almost a kilometer down into the twisty entrails of the earth to make art and to leave it there.

The discovery of the clay bison of the Tuc D'Audoubert caves was one of many shocking twentieth-century discoveries of highly sophisticated cave art stretching back tens of thousands of years.[2] The discoveries irreversibly changed our sense of what our Stone Age ancestors were like. They were not furry, grunting troglodytes. They had artistic souls. They showed us that humans are—by nature, not just by culture—art-making, art-consuming, art-addicted apes.

Prehistoric people lived short, hard, dangerous lives. They faced hostile tribes, dangerous beasts, and cruel winters. There were mates to woo, children to feed, rivals to defeat. Why did they brave those caves, with their bears and killer mazes and demon-haunted blackness? Why did they go there to paint and to sculpt, and maybe to sing, dance, and tell stories?

When it comes to the sculpture of the two bison, we will never know for sure. But story is now and always has been the queen of the arts. Our love of music, sculpture, painting, and dance are, in large part, different expressions of our infatuation with storytelling. When we go to a ballet, we do so to watch stories dance. And when we stroll a museum like the Louvre, we are strolling through a massive anthology of powerful stories from mythology and history, whether those stories are told in stone or paint or woven cloth. People go to the Louvre to marvel at the technical skill of the artists, yes, but also to experience, in visual art, Western culture's most durable tales of tragic heroes, sad maidens, and furious gods.

So, the smart money says the sculpture of the two bison is religious in character, reflecting a tribe's most precious stories of gods, spirits, origins, or endings.[3] And you can see, from footprints of adults and children running down the hardened mud of the corridors, that the sculptor's people came to marvel at his work and enter into his story.

There's something achingly beautiful about the humanity of all this. But it also suggests how deep our storytelling instincts run, and how difficult it will be to change *anything* about the ways people have always told and consumed stories. Even if you shape a tale in clay, then leave it entombed in the blackness of the earth, people will risk their lives—swimming, climbing, and crawling—to get to it.

Storytelling is, as I put it at the start, humanity's essential poison—as necessary for our survival as oxygen, and similarly destructive. I wrote this book partly in hopes of devising strategies for maintaining the essential, joyful side of storytelling while distilling out the worst of the poison. If I didn't think this was possible on slightly more days than not, I wouldn't have gone through the considerable bother of writing this book.

But as I think about the prehistoric sculptor cutting his story into clay, and as I scan back over the pages I've written, I'm beset by doubt. Let me put the case for pessimism as starkly as I can. It's hard to distill the poison out of story for a reason that now seems obvious to me in hindsight: it's already a purified distillate. To distill is to separate the different parts of a mixture. So distilled water is H_2O with all the impurities left behind. Liquor is ethanol with all yeasty carbohydrate mash left behind in the fermentation barrel. And a story is the intoxicant of the universal grammar boiled away from the bland mash of dull, disorderly, unstorified reality.

The only way to neutralize the toxic components of the essential poison would be to stir that bland mash back into the distillate—the boring bits, the ethically irrelevant bits, the unemotional bits, and all the good days protagonists spent basking in the sun. To imagine this possible would be to tumble back into Plato's first dream that humans might do without story altogether.

We can't. And even if we could, we'd never choose it, not even to save the whole world. We prefer the intoxication of distilled stories—even with all the potential for drunken mayhem—to the grimness of sobriety.

For as long as there have been humans, we've been telling the same old stories, in the same old way, for the same old reasons. And short of the raw repression that Plato recommends in *The Republic* and that the world's autocrats have always tried to impose, this will be hard to change. This is a frightening conclusion to reach because it's obvious that human beings can't live *without* stories. But, as technology continues to amplify their power, we may not be able to live *with* them much longer either.

Know Thyself

Plato's attack on story is famous. But I doubt it's ever changed any human being's daily relationship to stories. The philosopher Karl Popper gave Plato a share of blame for the crimes of the Nazis and the Soviets,[4] but parallels between Plato's ideal republic and the various totalitarian dystopias—many of them ruled by fantastic ignoramuses—would seem to be more the result of zealous utopian fantasists thinking alike than deep engagement with classical literature.

Aside from outraging poets and provoking their countermanifestos,[5] Plato's warnings about storytelling made so little impact for a simple reason: almost no one believed him. It's hard for storytelling animals to accept that the tales that give us so much goodness, meaning, and pleasure could also be the primary roots of our chaos, illogic, and cruelty.

And most of us still can't believe it. Early in my research for this book, I spent a morning in the lounge of the college psychology department, eagerly perusing the tables of contents and indexes of about twenty recent textbooks from different subfields of psychology. I was scanning for references to any variant of the word *story* or *narrative.* I had my notebook out to scribble down ideas, concepts, and references to journal articles to run down. When I finished, the notebook paper remained pristine. I got zero hits.

The science of stories exists, and this book couldn't have been written without it. But it's still a very young science, where the known is dwarfed by the unknown. And far from moving toward its rightful place near the heart of the human sciences, story science hasn't even penetrated the textbooks.

The Delphic admonition "know thyself" is at the foundation of intellectual life and all schemes of social improvement. We can't fix ourselves if we don't know ourselves. And I've argued that we don't know ourselves well enough. We don't know that we are just as much *Homo fictus* as *Homo sapiens*, and not knowing this is threatening everything.

If we hope to address civilization's largest problems, we need a much better sense for the sneaky ways that stories work on our minds and societies. This means encouraging scholars across the humanities and human sciences to embark on a massively interdisciplinary effort in the new field of narrative psychology, combining the thick, granular knowledge of the humanities with the special tools of the sciences.

Excuse me while I whisper in the ears of ambitious young researchers: *The tree of story science is heavy with toothsome, low-hanging fruit. Go feast yourself and grow fat in reputation. And while you're at it, why not help save the world to boot? This, brave scholar, is what's known as a call to adventure.*

As for the rest of us, just because we don't yet know enough about our storytelling minds doesn't mean we know nothing or that it's too soon to begin putting this knowledge to work.

When I shouted, at the start of our time together, "NEVER TRUST A STORYTELLER," I wasn't only talking *to* you but *about* you. We are all storytellers, and therefore not to be trusted—least of all by ourselves.

At an individual level, we all need to be more aware of the LARPish human tendency to enter into stories that are more invigorating and sharply drawn than reality and to then refuse to come out because the simulation is a better place to live than the boring and morally ambiguous real world. Each of us should try, in particular, to develop a personal suspicion not just about the moralistic simplifications of stories told by others but also about the ones we tell ourselves.

But it's so easy to smell the bad breath of others and so hard to whiff our own. I imagine my readers bent over these pages happily ticking off the ways my points apply to the people and narratives they don't like, while carving out exemptions for the narratives that are most dear to their own sense of a coherent world and identity. I'm not trying to insult you, reader. I'm just assuming you are human and, therefore, to one degree or another, lost in narrative.

But we aren't helpless. Once a person is told they have halitosis—and it usually takes a good friend to do it—they take more care with hygiene, they avoid problem foods, they stop close-talking their conversational partners. And perhaps they make a habit of cupping their hand over mouth and nose to check their own breath. Nothing to it!

And the same goes for the stories in our heads. We have to get into the habit of suspicion. We have to learn to sniff our own stories for exaggerations, fabrications, illogic, and other nonsense. This book is me telling you—as your friend—that you are almost certainly not exempt from the varieties of narrative halitosis covered in this book. No one is.

But I believe we can gain some control over our storytelling biases, just as we can learn to control other natural impulses. We just have to understand that these impulses are real, and that if they are left on autopilot, they will often lead us astray. For example, when I sense myself getting worked up into a fit of moralistic outrage—when I catch myself dehumanizing a person by turning them into a villain—I take a deep breath and try to imagine the story differently. In this way, I exert some sort of executive control over the automatic process of my brain. If I can't or won't do this, I'm not the master of the stories in my head, I'm just their slave, and I'm all the more degraded because I can't even sense the chains that hold me.

End Time

The golden age of Athens truly lasted only one generation, between the heroic war with Persia and the horrible decades of violence with Sparta. Everything we associate most strongly with ancient Athens, including vibrant democracy, the great imperial wealth that built the monuments of the Acropolis, and white-robed philosophers trying to chat their way to eternal truths—all of this was mainly a product of a bright flash of peace between the two great wars.

But Plato was not born in the age of Athenian splendor. He was born in the time of decline that followed. He was born into the greatest plague Athens had ever seen, and its longest and most brutal war. Far from unifying Athens, the war with Sparta inflamed every divide. Plato saw Athens throw out the Spartan occupiers and fall immediately to civil war. He saw an age of mob rule dawning that claimed many lives, including that of his beloved teacher, Socrates.

Nothing was ever solved. War seemed like a perpetual and incurable disease, with men doomed, as Homer put it, "to fight bloody wars from youth until we perish."[6] Within a decade after the end of the Peloponnesian War, Athens was back at war with Sparta.

There was a widely shared sense in Plato's time that even grimmer days might be at hand.[7] Maybe a great enemy would swarm over the horizon carrying spears and slave collars. And if Greece wasn't smashed from the outside, it would just bleed itself to death from within. And the terrible thing is that people could sense the catastrophe approaching, but no one could agree on how to stop it.

Plato would live to see the rising strength of the Macedonian empire, and the first probing attacks that preceded wholesale invasion. Soon after his death, the Macedonian phalanxes swept down from the north, and the glory that was Greece passed into history.

Much about Plato's historical circumstances is mirrored in ours. A raging pandemic. Wars that stretch over decades. The rise of remorseless demagogues leading populist movements. Ethnic and class tensions boiling over. Falling confidence in our civilization, rising foreign powers, and ever-more-plausible existential threats (in our case, nukes, climate change, novel plagues, the rise of AI, and old-fashioned tribal strife). We also share with Plato's age the dread of passing over into a post-truth existence where, thanks to sophistry of various kinds, people will stop seeing the same reality. And how can we unite to solve our problems if we can't agree which problems are real and which are just stories?

Plato didn't write *The Republic* for us. I doubt he imagined that young philosophy students would still be slogging their way through his book on its twenty-fourth centennial or that guys like me would be dissecting it. Plato wrote *The Republic* to save his world. He failed. He didn't know enough. He lived in a largely prescientific age and he had few giant shoulders to stand upon. But, if we heed Plato's warnings about the dangers of tale-telling—while trying to devise modern scientifically informed solutions—we might yet save ours.

The most important step is to develop more charitable rules of thumb for navigating the stories that divide us. I propose the following:

Hate and resist the story.
But try hard not to hate the storyteller.
And, for the sake of peace and your own soul, don't despise the poor sap who literally couldn't help falling for it.

Controlling the automatic ways our brains consume and create stories will be hard, and in the end we may fail. The storytelling instincts that helped build our species may turn and crush us. But if the danger weren't real, and the solutions weren't elusive, there'd be no need of heroes.

This, brave reader, is what's known as a call to adventure.

ACKNOWLEDGMENTS

THANKS TO THE ONES WHO HELPED. MY WIFE, TIFFANI. MY BROTHER, Garcia Roberto. My mother, Marcia. My friends Brian Boyd, Mathias Clasen, Tara Fee, Melanie Green, and Jennifer Harding. My production editor, Kelly Lenkevich. My copyeditor, Christina Palaia. My agent, Tom Miller. My editor, Eric Henney.

REFERENCES

Ahren, Raphael. 2020. "World War II Was Caused by Hatred of Jews, Preeminent Holocaust Scholar Says." *Times of Israel*, January 20, 2020. www.timesofisrael.com/world-war-ii-was-caused-by-hatred-of-jews-preeminent-holocaust-scholar-says/.

Allen, M., and R. W. Preiss. 1997. "Comparing the Persuasiveness of Narrative and Statistical Evidence Using Meta-Analysis." *Communication Research Reports* 14 (2): 125–131.

Alter, Alexandra. 2020. "The 'Trump Bump' for Books Has Been Significant. Can It Continue?" *New York Times*, December 24, 2020, B1.

Andersen, Marc Malmdorf, Uffe Schjoedt, Henry Price, Fernando E. Rosas, Coltan Scrivner, and Mathias Clasen. 2020. "Playing with Fear: A Field Study in Recreational Horror." *Psychological Science* 31 (12): 1497.

Anderson, Ross. 2020. "The Panopticon Is Already Here." *The Atlantic*, September 2020, 56–68.

Appel, Markus. 2008. "Fictional Narratives Cultivate Just-World Beliefs." *Journal of Communications* 58:62–83.

Appel, Markus, and Tobias Richter. 2007. "Persuasive Effects of Fictional Narratives Increase over Time." *Media Psychology* 10:113–135.

Arendt, Hannah. (1948) 1994. *The Origins of Totalitarianism.* Reprint, New York: Harcourt, Brace, Jovanovich.

Argo, Jennifer, Rui Zhu, and Darren W. Dahl. 2008. "Fact or Fiction: An Investigation of Empathy Differences in Response to Emotional Melodramatic Entertainment." *Journal of Consumer Research* 34:614–623.

Arieti, James. 1991. *Interpreting Plato: The Dialogues as Drama.* Lanham, MD: Rowman & Littlefield Publishers.

Asma, Stephen. 2017. *The Evolution of Imagination.* Chicago: University of Chicago Press.

Atran, Scott. 2003. "Genesis of Suicide Terrorism." *Science* 299:1534–1539.

———. 2006. "The Moral Logic and Growth of Suicide Terrorism." *Washington Quarterly* 29:127–1247.

Aubert, Maxime, Rustan Lebe, Adhi Agus Oktaviana, Muhammad Tang, Basran Burhan, Hamrullah, Andi Jusdi, et al. 2019. "Earliest Hunting Scene in Prehistoric Art." *Nature* 576:442–445.

AVAAZ. 2020. *Facebook's Algorithm: A Major Threat to Public Health.* August 19, 2020. avaazimages.avaaz.org/facebook_threat_health .pdf.

Azéma, Marc, and Florent Rivère. 2012. "Animation in Palaeolithic Art: A Pre-Echo of Cinema." *Antiquity* 86:316–324.

Bail, Christopher, Lisa P. Argyle, Taylor W. Brown, John P. Bumpus, Haohan Chen, M. B. Fallin Hunzaker, Jaemin Lee, Marcus Mann, Friedolin Merhout, and Alexander Volfovsky. 2018. "Exposure to Opposing Views Can Increase Political Polarization: Evidence from a Large-Scale Field Experiment on Social Media." *Proceedings of the National Academy of Sciences* 115 (37): 9216–9221

Bailey, Blake. 2004. *A Tragic Honesty: The Life and Work of Richard Yates.* New York: Picador.

Baird, Jay. 1974. *The Mythical World of Nazi War Propaganda 1939–1945.* Minneapolis: University of Minnesota Press.

Baldwin, James. (1963) 1992. *The Fire Next Time.* New York: Vintage.

Barnes, Jennifer, and Paul Bloom. 2014. "Children's Preference for Social Stories." *Developmental Psychology* 50:498–503.

Barraza, Jorge, Veronika Alexander, Laura E. Beavin, Elizabeth T. Terris, and Paul J. Zak. 2015. "The Heart of the Story: Peripheral Physiology During Narrative Exposure Predicts Charitable Giving." *Biological Psychology* 105:138–143.

Baumeister, Roy, Ellen Bratslavsky, Catrin Finkenauer, and K. de Vohs. 2001. "Bad Is Stronger Than Good." *Review of General Psychology* 5:323–370.

Bazelon, Emily. 2020. "The Problem of Free Speech in an Age of Disinformation." *New York Times Magazine*, October 13, 2020, 26.

Bedard, Paul. 2018. "Trump Bump for Media Doubles, Drives Millions to News Websites." *Washington Examiner*, June 14, 2018. www.washingtonexaminer.com/washington-secrets/trump-bump-for-media-doubles-drives-millions-to-news-websites.

Begouen, Max. 1926. *Bison of Clay*. Translated by Robert Luther Duffus. Harlow, UK: Longmans, Green & Co.

Begouen, Robert, Carole Fritz, Gilles Tosello, Andreas Pastoors, François Faist, and Jean Clottes. 2009. *Le Sanctuaire Secret des Bisons: Il Y A 14,000 Ans, Dans la Caverne Du Tuc DAudobert.* Somogy Editions D'Art.

Benson, Hugh, ed. 2006. *A Companion to Plato.* Hoboken, NJ: Blackwell.

Bentham, Jeremy. 1791. *Panopticon; or, The Inspection House.* London: T. Payne.

Berger, Jonah. 2013. *Contagious: Why Things Catch On.* New York: Simon & Schuster.

Berger, Jonah, and Katherine Milkman. 2012. "What Makes Online Content Viral." *Journal of Marketing Research* 49 (2): 192–205.

Bezdek, Matthew, Jeffrey Foy, and Richard J. Gerrig. 2013. "'Run for It!': Viewers' Participatory Responses to Film Narratives." *Psychology of Aesthetics, Creativity, and the Arts* 7:409–416.

Bezdek, Matthew, and Richard Gerrig. 2017. "When Narrative Transportation Narrows Attention: Changes in Attentional Focus During Suspenseful Film Viewing." *Media Psychology* 20:60–89.

Bietti, Lucas M., Ottilie Tilston, and Adrian Bangerter. 2018. "Storytelling as Adaptive Collective Sensemaking." *Topics in Cognitive Science* 11:1–23.

Blondell, Ruby. 2002. *The Play of Character in Plato's Dialogues.* Cambridge: Cambridge University Press.

Bloom, Allan. 1968. "Interpretive Essay." In *Republic of Plato,* trans. Allan Bloom. New York: Basic Books.

Bloom, Harold. 1994. *The Western Canon: The Books and School of the Ages.* New York: Harcourt Brace & Company.

Bloom, Paul. 2016. *Against Empathy: The Case for Rational Compassion.* New York: HarperCollins.

Boehm, Christopher. 2001. *Hierarchy in the Forest: The Evolution of Egalitarian Behavior.* Cambridge, MA: Harvard University Press.

Bohannon, Laura. 1966. "Shakespeare in the Bush." *Natural History,* August–September 1966, 28–33.

Bower, Bruce. 2019. "A Nearly 44,000-Year-Old Hunting Scene Is the Oldest Known Storytelling Art." *ScienceNews,* December 11, 2019. www.sciencenews.org/article/nearly-44000-year-old-hunting-scene-is-oldest-storytelling-art.

Bower, G. H., and M. C. Clark. 1969. "Narrative Stories as Mediators for Serial Learning." *Psychonomic Science* 14:181–182.

Bowker, Gordon. 2011. *James Joyce: A New Biography.* New York: Farrar, Straus and Giroux.

Boyd, Brian. 2009. *On the Origin of Stories: Evolution, Cognition, Fiction.* Cambridge, MA: Harvard University Press.

Boyd, Brian, Joseph Carroll, and Jonathan Gottschall. 2010. *Evolution, Literature, and Film: A Reader.* New York: Columbia University Press.

Boyer, Pascal. *Religion Explained: The Evolutionary Origins of Religious Thought.* New York: Basic Books, 2002.

Bracken, Bethany, Veronika Alexander, Paul J. Zak, Victoria Romero, and Jorge A. Barraza. 2014. "Physiological Synchronization Is Associated with Narrative Emotionality and Subsequent Behavioral Response." In *Augmented Cognition,* ed. D. Schmorrow and C. Fidopiastis, 3–13. New York: Springer.

Braddock, Kurt, and James Dillard. 2016. "Meta-Analytic Evidence for the Persuasive Effect of Narratives on Beliefs Attitudes, Intentions, and Behaviors." *Communications Monographs* 83:446–467.

Brady, William, Julian A. Wills, John T. Jost, Joshua A. Tucker, and Jay J. Van Bavel. 2017. "Emotion Shapes the Diffusion of Moralized Content in Social Networks." *PNAS* 114:7313–7318.

Branch, Glenn, and Craig Foster. 2018. "Yes, Flat-Earthers Really Do Exist." *Scientific American,* October 24, 2018. blogs.scientificamerican.com/observations/yes-flat-earthers-really-do-exist/.

Brandenberg, David. 2011. *Propaganda State in Crisis.* New Haven, CT: Yale University Press.

Brechman, Jean Marie, and Scott Purvis. 2015. "Narrative Transportation and Advertising." *International Journal of Advertising* 34:366-381.

Breithaupt, Fritz. 2019. *The Dark Sides of Empathy.* Ithaca, NY: Cornell University Press.

Brenan, Megan. 2020. "Americans Remain Distrustful of Mass Media." *Gallup*, September 30, 2020. news.gallup.com/poll/321116/americans-remain-distrustful-mass-media.aspx.

Breuil, Abbé Henri. 1979. *Four Hundred Centuries of Cave Art.* Hacker Art Books.

Brinthaupt, Thomas M. 2019. "Individual Differences in Self-Talk Frequency: Social Isolation and Cognitive Disruption." *Frontiers in Psychology* 10. doi:10.3389/fpsyg.2019.01088.

Brodrick, Alan Houghton. 1963. *Father of Prehistory: The Abbé Henri Breuil: His Life and Times.* New York: William Morrow.

Brown, Donald. 1991. *Human Universals.* New York: McGraw-Hill.

Burdick, Alan. 2018. "Looking for Life on a Flat Earth." *New Yorker.* May 30, 2018. www.newyorker.com/science/elements/looking-for-life-on-a-flat-earth.

Burke, D., and N. Farbman. 1947. "The Bushmen: An Ancient Race Struggles to Survive in the South African Deserts." *Life*, February 3, 1947, 91–99.

Burnard, Trevor. 2011. "The Atlantic Slave Trade." In *The Routledge History of Slavery*, edited by Gad Heuman and Trevor Burnard, 80–98. London: Routledge.

Burroway, Janet. 2003. *Writing Fiction: A Guide to Narrative Craft.* 6th ed. New York: Longman.

Campbell, Joseph. 1949. *The Hero with a Thousand Faces.* 1st ed. Princeton, NJ: Princeton University Press.

Cantor, Joan. 2009. "Fright Reactions to Mass Media." In *Media Effects: Advances in Theory and Research*, edited by Jennings Bryant and Mary Beth Oliver, 287–306. New York: Routledge.

Carroll, Joseph, Jonathan Gottschall, Dan Kruger, and John Johnson. 2012. *Graphing Jane Austen: The Evolutionary Basis of Literary Meaning.* New York: Palgrave Macmillan.

Cha, Ariana Eunjung. 2015. "Why DARPA Is Paying People to Watch Alfred Hitchcock Cliffhangers." *Washington Post*, July 28, 2015. www.washingtonpost.com/news/to-your-health/wp/2015/07/28/why-darpa-is-paying-people-to-watch-alfred-hitchcock-cliffhangers/.

Chang, Edward, Katherine L. Milkman, Laura J. Zarrow, Kasandra Brabaw, Dena M. Gromet, Reb Rebele, Cade Massey, Angela L. Duckworth, and Adam Grant. 2019. "Does Diversity Training Work the Way It's Supposed To?" *Harvard Business Review*,

July 9, 2019. hbr.org/2019/07/does-diversity-training-work-the-way-its-supposed-to.

Chen, Tsai. 2015. "The Persuasive Effectiveness of Mini-Films: Narrative Transportation and Fantasy Proneness." *Journal of Consumer Behavior* 14:21–27.

Chiu, Melissa, and Zheng Shengtian, eds. 2008. *Art and China's Revolution.* New Haven, CT: Yale University Press.

Chomsky, Noam. 1965. *Aspects of the Theory of Syntax.* Cambridge, MA: MIT Press.

Chua, Amy. 2007. *Day of Empire: How Hyperpowers Rise to Global Dominance—and Why They Fail.* New York: Anchor Books.

———. 2018. *Political Tribes.* New York: Penguin.

Cillizza, Chris. 2014. "Just 7 Percent of Journalists Are Republicans. That's Far Fewer Than Even a Decade Ago." *Washington Post*, May 6, 2014. www.washingtonpost.com/news/the-fix/wp/2014/05/06/just-7-percent-of-journalists-are-republicans-thats-far-less-than-even-a-decade-ago/.

Clark, Daniel, director. 2018. *Behind the Curve* (film). Netflix.

Cohen, Anna-Lisa, E. Shavalian, and M. Rube. 2015. "The Power of the Picture: How Narrative Film Captures Attention and Disrupts Goal Pursuit." *PLOS ONE* 10 (12). doi.org/10.1371/journal.pone.0144493.

Colapinto, John. 2021. *This Is the Voice.* New York: Simon & Schuster.

Coleridge, Samuel Taylor. 1817. *Biographia Literaria.* London: Rest Fenner.

Copeland, Libby. 2017. "Why Mind Wandering Can Be So Miserable, According to Happiness Experts." *Smithsonian Magazine*, February 24, 2017. www.smithsonianmag.com/science-nature/why-mind-wandering-can-be-so-miserable-according-happiness-experts-180962265/.

Corballis, Michael. 2015. *The Wandering Mind: What the Brain Does When You're Not Looking.* Chicago: University of Chicago Press.

Corman, Steven, Adam Cohen, Anthony Roberto, Gene Brewer, and Scott Ruston. 2013. "Toward Narrative Disruptors and Inductors: Mapping the Narrative Comprehension Network and Its Persuasive Effects" (research project). Arizona State University. asu.pure.elsevier.com/en/projects/toward-narrative-disruptors-and-inductors-mapping-the-narrative-c-10.

Correa, Kelly, Bradly T. Stone, Maja Stikic, Robin R. Johnson, and Chris Berka. 2015. "Characterizing Donation Behavior from Psychophysiological Indices of Narrative Experience." *Frontiers in Neuroscience* 9:1–15.

Crandall, Kelly. 2006. "Invisible Commercials and Hidden Persuaders: James M. Vicary and the Subliminal Advertising Controversy of 1957" (undergraduate honors thesis, University of Florida). plaza .ufl.edu/cyllek/docs/KCrandall_Thesis2006.pdf.

Crispin Miller, Mark. 2007. *Introduction to Packard, Vance. The Hidden Persuaders.* New York: Ig Publishing.

Cunliffe, Richard. 1963. *A Lexicon of the Homeric Dialect.* Norman: University of Oklahoma Press.

Dabrowska, Ewa. 2015. "What Exactly Is Universal Grammar, and Has Anyone Seen It?" *Frontiers in Psychology* 6. doi.org/10.3389 /fpsyg.2015.00852.

Dahlstrom, Michael. 2014. "Using Narratives and Storytelling to Communicate Science with Nonexpert Audiences." *Proceedings of the National Academy of Sciences* 111:13614–13620.

Dal Cin, Sonya, Mike Stoolmiller, and James D. Sargent. 2012. "When Movies Matter: Exposure to Smoking in Movies and Changes in Smoking Behavior." *Journal of Health Communication* 17 (1): 76–89.

Damasio, Antonio. 2005. *Descartes' Error: Emotion, Reason, and the Human Brain.* London: Penguin.

———. 2010. *Self Comes to Mind: Constructing the Conscious Brain.* New York: Pantheon.

Darwin, Charles. 1871. *The Descent of Man and Selection in Relation to Sex.* New York: D. Appleton and Company.

Davies, Joanna, Josiane Cillard, Bertrand Friguet, Enrique Cadenas, Jean Cadet, Rachael Cayce, Andrew Fishmann, et al. 2017. "The Oxygen Paradox, the French Paradox and Age-Related Diseases." *Geroscience* 39:499–550.

Davies, Kelvin J. A. 2016. "The Oxygen Paradox, Oxidative Stress, and Ageing." *Archives of Biochemistry and Biophysics* 595:28–32.

Davies, Kelvin J. A., and Fulvio Ursini. 1995. *The Oxygen Paradox.* Padua, Italy: CLEUP University Press.

Dawkins, Richard. 2008. *The God Delusion.* Boston: Mariner Books.

Defense Advanced Research Projects Agency. n.d. "Narrative Networks (Archived)." www.darpa.mil/program/narrative-networks.

De Graaf, Anneke, and Lettica Histinx. 2011. "The Effect of Story Structure on Emotion, Transportation, and Persuasion." *Information Design Journal* 19:142–154.

Dehghani, Morteza, Reihane Boghrati, Kingson Man, Joe Hoover, Sarah I. Gimbel, Ashish Vaswani, Jason D. Zevin, et al. 2017. "Decoding the Neural Representation of Story Meanings Across Languages." *Human Brain Mapping* 38 (12): 6096–6106. doi:10.1002/hbm.23814.

Del Giudice, Marco, Tom Booth, and Paul Irwing. 2012. "The Distance Between Mars and Venus: Measuring Global Sex Differences in Personality." *PLOS ONE* 7 (1). doi.org/10.1371/journal.pone.0029265.

Dennett, Daniel. 2006. *Breaking the Spell: Religion as a Natural Phenomenon.* New York: Penguin.

Dibble, J. L., and S. F. Rosaen. 2011. "Parasocial Interaction as More Than Friendship." *Journal of Media Psychology* 23:122–132.

Didion, Joan. 1976. "Why I Write." *New York Times Book Review,* December 5, 1976, 270. www.nytimes.com/1976/12/05/archives/why-i-write-why-i-write.html.

Dill-Shackleford, Karen E., and Cynthia Vinney. 2020. *Finding Truth in Fiction: What Fan Culture Gets Right—and Why It's Good to Get Lost in a Story.* New York: Oxford University Press.

Dines, Gail. 2011. *Pornland: How Porn Has Hijacked Our Sexuality.* Boston: Beacon Press.

Dissanayake, Ellen. 1990. *What Is Art For?* Seattle: University of Washington Press.

———. 1995. *Homo Aetheticus.* Seattle: University of Washington Press.

Djikic, Maja, and Keith Oatley. 2013. "Reading Other Minds: Effects of Literature on Empathy." *Scientific Study of Literature* 3:28–47.

Dobbin, Frank, and Alexandra Kale. 2013. "The Origins and Effects of Corporate Diversity Programs." In *Oxford Handbook of Diversity and Work,* edited by Quinetta M. Roberson, 253–281. Oxford University Press.

Dor, Daniel. 2015. *The Instruction of Imagination: Language as a Social Communication Technology.* New York: Oxford University Press.

Dreier, Peter. 2017. "Most Americans Are Liberal, Even If They Don't Know It." *American Prospect,* November 10, 2017. prospect.org/article/most-americans-are-liberal-even-if-they-don%E2%80%99t-know-it.

Drum, Kevin. 2018. "America Is Getting More Liberal Every Year." *Mother Jones,* January 22, 2018. www.motherjones.com/kevin-drum/2018/01/america-is-getting-more-liberal-every-year/.

Duffy, Bobby. 2017. "Opinion. Crime, Terrorism and Teen Pregnancies: Is It All Doom and Gloom? Only in Our Minds." *The Guardian*, December 8, 2017.

Dunbar, Robin, Ben Teasdale, Jackie Thompson, Felix Budelmann, Sophie Duncan, Evert van Emde Boas, and Laurie Maguire. 2016. "Emotional Arousal When Watching Drama Increases Pain Threshold and Social Bonding." *Royal Society Open Science* 3 (9). doi.org/10.1098/rsos.160288.

Dutton, Denis. 2009. *The Art Instinct: Beauty, Pleasure, and Human Evolution.* London: Bloomsbury Press.

Ecker, Ullruch K. H., Lucy H. Butler, and Anne Hamby. 2020. "You Don't Have to Tell a Story! A Registered Report Testing the Effectiveness of Narrative Versus Non-Narrative Misinformation Corrections." *Cognitive Research: Principles and Implications* 5 (64). doi.org/10.1186/s41235-020-00266-x.

Ehrenreich, John. 2021. "Why People Believe in Conspiracy Theories." *Slate*, January 11, 2021. slate.com/technology/2021/01/conspiracy-theories-coronavirus-fake-psychology.html.

Ehrman, Bart. 2007. *Misquoting Jesus: The Story Behind Who Changed the Bible and Why.* New York: HarperOne.

———. 2014. *How Jesus Became God: The Exaltation of a Jewish Preacher from Galilee.* New York: Harper One.

———. 2018. *The Triumph of Christianity: How a Forbidden Religion Swept the World.* New York: Simon & Schuster.

———. 2020. *Heaven and Hell: A History of the Afterlife.* New York: Simon & Schuster.

Ellis, Christopher, and James Stimson. 2012. *Ideology in America.* New York: Cambridge University Press.

Ellithorpe, Morgan, and Sarah Brookes. 2018. "I Didn't See That Coming: Spoilers, Fan Theories, and Their Influence on Enjoyment and Parasocial Breakup Distress During a Series Finale." *Psychology of Popular Media Culture* 7:250–263.

Emerson, Ralph Waldo. 1850. *Representative Men: Seven Lectures.* London: George Routledge.

Enten, Harry. 2017. "Most People Believe in JFK Conspiracy Theories." *FiveThirtyEight*, October 23, 2017. fivethirtyeight.com/features/the-one-thing-in-politics-most-americans-believe-in-jfk-conspiracies/.

Fessler, Daniel, Anne C. Pisor, and Carlos David Navarrete. 2014. "Negatively-Biased Credulity and the Cultural Evolution of Beliefs." *PLOS ONE* 9. doi.org/10.1371/journal.pone.0095167.

Fields, Douglas. 2020. "Mind Reading and Mind Control Technologies Are Coming." Observations (blog), *Scientific American*, March 10, 2020. blogs.scientificamerican.com/observations/mind-reading-and-mind-control-technologies-are-coming/.

Flavel, John H., Eleanor R. Flavell, Frances L. Green, and Jon E. Korfmacher. 1990. "Do Young Children Think of Television Images as Pictures or Real Objects?" *Journal of Broadcasting and Electronic Media* 34:399–419.

Flesch, William. 2007. *Comeuppance: Costly Signaling, Altruistic Punishment, and Other Biological Components of Fiction.* Cambridge, MA: Harvard University Press.

Foer, Joshua. 2012. *Moonwalking with Einstein: The Art and Science of Remembering Everything.* New York: Penguin.

Forster, E. M. (1927) 1956. *Aspects of the Novel.* Boston: Mariner Books.

Franks, Bradley, Adrian Bangerter, and Martin Bauer. 2013. "Conspiracy Theories as Quasi-Religious Mentality: An Integrated Account from Cognitive Science, Social Representations Theory, and Frame Theory." *Frontiers in Psychology* 4:1–424.

Frontline. 2020. "China Undercover." Season 2020, episode 9. Aired April 7, 2020. www.pbs.org/video/china-undercover-zqcoh2/.

Frye, Northrop. (1957) 2020. *The Anatomy of Criticism.* Princeton, NJ: Princeton University Press.

Fukuyama, Francis. 1992. *The End of History and the Last Man.* New York: Free Press.

Gallo, Carmine. 2016. *The Storyteller's Secret: Why Some Ideas Catch On and Others Don't.* New York: St. Martin's.

Gardner, John. 1978. *On Moral Fiction.* New York: Basic Books.

———. 1983. *The Art of Fiction: Notes on Craft for Young Writers.* New York: Vintage.

Garwood, Christine. 2007. *Flat Earth: The History of an Infamous Idea.* New York: Macmillan.

Gates, Henry Louis. 1999. *Wonders of the African World.* New York: Knopf.

———. 2010. "Ending the Slavery Blame-Game." *New York Times,* April 23, 2010, A27.

Gerbner, George, Larry Gross, Michael Morgan, and Nancy Signorielli. 2006. "The 'Mainstreaming' of America: Violence Profile No. 11." *Journal of Communication* 30:10–29.

Gerrig, Richard. 1993. *Experiencing Narrative Worlds: On the Psychological Activities of Reading*. New Haven, CT: Yale University Press.

Geurts, Bart. 2018. "Making Sense of Self Talk." *Review of Philosophy and Psychology* 9:271–285.

Ghose, Tia. 2016. "Half of Americans Believe in 9/11 Conspiracy Theories." *Live Science*, October 13, 2016. www.livescience.com/56479-americans-believe-conspiracy-theories.html.

Giles, David. 2002. "Parasocial Interaction: A Review of the Literature and a Model for Future Research." *Media Psychology* 4:279–305.

Godin, Seth. 2012. *All Marketers Tell Stories*. New York: Portfolio.

Goldberg, Jeffrey. 2020. "Why Obama Fears for Our Democracy." *The Atlantic*, November 16, 2020. www.theatlantic.com/ideas/archive/2020/11/why-obama-fears-for-our-democracy/617087/.

Gottschall, Jonathan. 2008. *Literature, Science, and a New Humanities*. New York: Palgrave Macmillan.

———. 2012. *The Storytelling Animal: How Stories Make Us Human*. New York: Houghton Mifflin Harcourt.

———. 2013. "Story 2.0: The Surprising Thing About the Next Wave of Narrative." *Fast Company*. October 27, 2013. www.fastcompany.com/3020047/story-20-the-surprising-thing-about-the-next-wave-of-narrative.

———. 2016. "Afterword." *Darwin's Bridge: Uniting the Humanities and Sciences*, edited by Joseph Carroll et al., 269–272. New York: Oxford University Press.

Gould, Stephen Jay. 1994. *Eight Little Piggies: Reflections in Natural History*. New York: Norton.

Gourevitch, Philip. 1998. *We Wish to Inform You That Tomorrow We Will Be Killed with Our Families: Stories from Rwanda*. New York: Picador.

Grabe, Maria Elizabeth. 2012. "News as Reality-Inducing, Survival-Relevant, and Gender-Specific Stimuli." In *Applied Evolutionary Psychology*, edited by S. Craig Roberts, 361–377. New York: Oxford University Press.

Graesser, A. C., K. Hauft-Smith, A. D. Cohen, and L. D. Pyles. 1980. "Advanced Outlines, Familiarity, Text Genre, and Retention of Prose." *Journal of Experimental Education* 48:209–220.

Graesser, A. C., N. L. Hoffman, and L. F. Clark. 1980. "Structural Components of Reading Time." *Journal of Verbal Learning and Verbal Behavior* 19 (2): 135–151.

Green, Melanie. 2008. "Research Challenges in Narrative Persuasion." *Information Design Journal* 6:47–52.

Green, Melanie, and Timothy Brock. 2000. "The Role of Transportation in the Persuasiveness of Public Narratives." *Journal of Personality and Social Psychology* 79:701–721.

Green, Melanie, and Jenna Clark. 2012. "Transportation into Narrative Worlds: Implications for Entertainment Media Influences on Tobacco Use." *Addiction* 108:477–484.

Green, Melanie, and K. E. Dill. 2013. "Engaging with Stories and Characters: Learning, Persuasion, and Transportation into Narrative Worlds." In *Oxford Handbook of Media Psychology*, edited by K. E. Dill, 449–461. New York: Oxford University Press.

Grube, G. M. A. 1927. "The Marriage Laws in Plato's Republic." *Classical Quarterly* 21 (2): 95–99.

Guber, Peter. 2011. *Tell to Win.* New York: Currency.

Guriev, Sergei, and Daniel Treisman. 2019. "Informational Autocrats." *Journal of Economic Perspectives* 33 (4): 100–127.

Haberstroh, Jack. 1994. *Ice Cube Sex: The Truth About Subliminal Advertising.* Notre Dame, IN: Cross Cultural Publications.

Haidt, Jonathan. 2012a. "Born This Way? Nature, Nurture, Narratives, and the Making of Our Political Personalities." *Reason*, May 2012. reason.com/2012/04/10/born-this-way/.

———. 2012b. *The Righteous Mind: Why Good People Are Divided by Politics and Religion.* New York: Pantheon.

Hakemulder, J. 2000. *The Moral Laboratory: Experiments Examining the Effects of Reading Literature on Social Perception and Moral Self-Concept.* Amsterdam: John Benjamins.

Hall, Alice. 2019. "Identification and Parasocial Relationships with Characters from *Star Wars: The Force Awakens.*" *Psychology of Popular Media Culture* 8 (1): 88–98.

Halper, Stefan. 2013. *China: The Three Warfares* (report). US Department of Defense. cryptome.org/2014/06/prc-three-wars.pdf.

Hambrick, David, and Alexander Burgoyne. 2016. "The Difference Between Rationality and Intelligence." *New York Times*, September 16, 2016, SR12.

Hamby, Ann, and David Brinberg. 2016. "Happily Ever After: How Ending Valence Influences Narrative Persuasion in Cautionary Stories." *Journal of Advertising* 45:498–508.

Hamby, Anne, David Brinberg, and Kim Daniloski. 2017. "Reflecting on the Journey: Mechanisms in Narrative Persuasion." *Journal of Consumer Psychology* 27:11–22.

Hamilton, Edith. 1961. Introduction. In *The Collected Dialogues of Plato.* Princeton, NJ: Princeton University Press.

Harari, Yuval Noah. 2015. *Sapiens: A Brief History of Humanity.* New York: Harper.

Harris, Sam. 2012. *Free Will.* New York: Free Press.

Hartman, R. J. 2019. "Moral Luck and the Unfairness of Morality." *Philosophical Studies* 176 (12): 3179–3197.

Havelock, Eric. 1963. *Preface to Plato.* Cambridge, MA: Harvard University Press.

Hayman, David. 1990. *The "Wake" in Transit.* Ithaca, NY: Cornell University Press.

Heath, C., C. Bell, and E. Sternberg. 2001. "Emotional Selection in Memes: The Case of Urban Legends." *Journal of Personality and Social Psychology* 81:1028–1041.

Heatherton, Todd, and James Sargent. 2009. "Does Watching Smoking in Movies Promote Teenage Smoking?" *Current Directions in Psychological Science* 18:63–67.

Heider, Fritz. 1983. *Life of a Psychologist: An Autobiography.* Lawrence: University of Kansas Press.

Heider, Fritz, and Marianne Simmel. 1944. "An Experimental Study of Apparent Behavior." *American Journal of Psychology* 57:243–259.

Henley, Jon, and Niamh McIntyre. 2020. "Survey Uncovers Widespread Belief in 'Dangerous' Covid Conspiracy Theories." *The Guardian,* October 26, 2020. www.theguardian.com/world/2020/oct/26/survey-uncovers-widespread-belief-dangerous-covid-conspiracy-theories.

Heuman, Gad, and Trevor Burnard. 2011. "Introduction." In *The Routledge History of Slavery,* edited by Gad Heuman and Trevor Burnard, 1–16. London: Routledge.

———, eds. 2011. *The Routledge History of Slavery.* London: Routledge.

Hill, Kashmir, and Jeremy White. 2020. "Do These A.I. Created Fake People Look Real to You?" *New York Times,* November 21, 2020.

www.nytimes.com/interactive/2020/11/21/science/artificial -intelligence-fake-people-faces.html.

Hobson, Katherine. 2018. "Clicking: How Our Brains Are in Sync." *Princeton Alumni Weekly*, April 11, 2018.

Hoeken, Hans, Matthijs Kolthoff, and José Sanders. 2016. "Story Perspective and Character Similarity as Drivers of Identification and Narrative Persuasion." *Human Communication Research* 42:292–311.

Hoeken, Hans, and Jop Sinkeldam. 2014. "The Role of Identification and Perception of Just Outcome in Evoking Emotions in Narrative Persuasion." *Journal of Communication* 64:935–955.

Hoffner, C. A., and E. L. Cohen. "Responses to Obsessive Compulsive Disorder on Monk Among Series Fans: Parasocial Relations, Presumed Media Influence, and Behavioral Outcomes." *Journal of Broadcasting & Electronic Media* 56 (4): 650–668.

Hogan, Patrick. 2003. *The Mind and Its Stories*. New York: Cambridge University Press.

Holmes, Marcia. 2017. "Edward Hunter and the Origins of 'Brainwashing'" (blog), Hidden Persuaders, May 25, 2017. www.bbk.ac.uk /hiddenpersuaders/blog/hunter-origins-of-brainwashing/.

Homer. 1999. *The Iliad*. Volume II. Edited by G. P. Goold. Cambridge, MA: Harvard University Press.

Howland, Jacob. 1993. *The Republic: The Odyssey of Philosophy*. Philadelphia: Paul Dry Books.

Hunt, Lynn. 2007. *Inventing Human Rights: A History*. New York: Norton.

Jacobsen, Annie. 2015. *The Pentagon's Brain: An Uncensored History of DARPA, America's Top Secret Military Research Agency*. Boston: Little, Brown.

James, Steven. 2018. "What a Coincidence: 7 Clever Strategies for Harnessing Coincidences in Fiction." *Writer's Digest*, September 7, 2018. www.writersdigest.com/write-better-fiction/what-a -coincidence-7-strategies-for-creating-clever-coincidences-in-fiction.

Janaway, Christopher. 1995. *Images of Excellence: Plato's Critique of the Arts*. New York: Oxford University Press.

———. 2006. "Plato and the Arts." In *A Companion to Plato*, edited by Hugh Benson, 388–400. Hoboken, NJ: Blackwell.

Jarrett, Christian. 2016. "Do Men and Women Really Have Different Personalities?" *BBC*, October 12, 2016. www.bbc.com/future/story/20161011-do-men-and-women-really-have-different-personalities.

Jaschik, Scott. 2016. "Professors, Politics and New England." *Inside Higher Ed*, July 5, 2016. www.insidehighered.com/news/2016/07/05/new-analysis-new-england-colleges-responsible-left-leaning-professoriate.

———. 2017. "Professors and Politics: What the Research Says." *Inside Higher Ed*, February 27, 2017. www.insidehighered.com/news/2017/02/27/research-confirms-professors-lean-left-questions-assumptions-about-what-means.

———. 2018. "Falling Confidence in Higher Ed." *Inside Higher Ed*, October 9, 2018. www.insidehighered.com/news/2018/10/09/gallup-survey-finds-falling-confidence-higher-education.

Johnson, Dan, Grace K. Cushman, Lauren A. Borden, and Madison S. McCune. 2013. "Potentiating Empathic Growth: Generating Imagery While Reading Fiction Increases Empathy and Prosocial Behavior." *Psychology of Aesthetics, Creativity, and the Arts* 7 (3): 306–312.

Jones, Kerry, Kelsey Libert, and Kristin Tynski. 2016. "The Emotional Combinations That Make Stories Go Viral." *Harvard Business Review*, May 23, 2016. hbr.org/2016/05/research-the-link-between-feeling-in-control-and-viral-content.

Jones, Jeffrey. 2018. "Confidence in Higher Education Down Since 2015." Gallup Blog (blog), Gallup, October 9, 2018. news.gallup.com/opinion/gallup/242441/confidence-higher-education-down-2015.aspx.

Kahn, Chris. 2020. "Half of Republicans Say Biden Won Because of a 'Rigged' Election: Reuters/Ipsos Poll." Reuters, November 18, 2020. www.reuters.com/news/picture/half-of-republicans-say-biden-won-becaus-idUSKBN27Y1AJ.

Kahneman, Daniel. 2011. *Thinking Fast and Slow.* New York: Farrar, Straus and Giroux.

Kang, Olivia, and Thalia Wheatley. 2017. "Pupil Dilation Patterns Spontaneously Synchronize Across Individuals During Shared Attention." *Journal of Experimental Psychology: General* 146:569–576.

Kenez, Peter. 1974. *The Birth of the Propaganda State: Soviet Methods of Mass Mobilization 1917–1929.* Cambridge: Cambridge University Press.

Kenny, Anthony. 2013. *Introduction to the Poetics of Aristotle.* Translated by Anthony Kenny. New York: Oxford University Press.

Kessler, Glenn. 2021. "Trump Made 30,573 False or Misleading Claims as President. Nearly Half Came in His Final Year." *Washington Post,* January 23, 2021. www.washingtonpost.com/politics/how-fact-checker-tracked-trump-claims/2021/01/23/ad04b69a-5c1d-11eb-a976-bad6431e03e2_story.html.

Key, Wilson Bryan. 1973. *Subliminal Seduction: Ad Media's Manipulation of a Not So Innocent America.* New York: Penguin.

Kidd, Daniel C., and Emanuele Castano. 2013. "Reading Literary Fiction Improves Theory of Mind." *Science,* October 18, 2013, 377–380.

Killingsworth, Matthew, and Daniel Gilbert. 2010. "A Wandering Mind Is an Unhappy Mind." *Science* 330 (November 12, 2010): 932.

Kim, Catherine. 2020. "Poll: 70 Percent of Republicans Don't Think the Election Was Free and Fair." *Politico,* November 11, 2020. www.politico.com/news/2020/11/09/republicans-free-fair-elections-435488.

Kinzer, Stephen. 2020. *Poisoner in Chief.* New York: St. Martin's.

Kirsch, Adam. 1968. Introduction. In *The Republic of Plato,* translated by Alan Bloom. New York: Basic Books.

Kjeldgaard-Christiansen, Jens. 2016. "Evil Origins: A Darwinian Genealogy of the Popcultural Villain." *Evolutionary Behavioral Sciences* 10:109–122.

Klein, Ezra. 2020. *Why We're Polarized.* New York: Simon & Schuster.

Klin, A. 2000. "Attributing Social Meaning to Ambiguous Visual Stimuli in Higher-Functioning Autism and Asperger Syndrome: The Social Attribution Task." *Journal of Child Psychology and Psychiatry* 41:831–846.

Kolbert, Elizabeth. 2017. "Why Facts Don't Change Our Minds." *New Yorker,* February 20, 2017. www.newyorker.com/magazine/2017/02/27/why-facts-dont-change-our-minds.

Krendl, A., C. Neil Macrae, William M. Kelley, Jonathan A. Fugelsang, and Todd F. Heatherton. 2006. "The Good, the Bad, and the Ugly: An FMRI Investigation of the Functional Anatomic Correlates of Stigma." *Social Neuroscience* 1:5–15.

Kroeger, Brooke. 2017. *The Suffragents: How Women Used Men to Get the Vote.* Albany: SUNY Press.

Kross, Ethan. 2021. *Chatter: The Voice in Our Head, Why It Matters, and How to Harness It.* New York: Crown.

Kuzmičová, A., A. Mangen, H. Støle, and A. C. Begnum. 2017. "Literature and Readers' Empathy: A Qualitative Text Manipulation Study." *Language and Literature* 26:137–152.

LaFrance, Adrienne. 2020. "The Prophecies of Q." *The Atlantic,* June 2020. www.theatlantic.com/magazine/archive/2020/06/qanon-nothing-can-stop-what-is-coming/610567/.

Langbert, Mitchell. 2019. "Homogenous: The Political Affiliations of Elite Liberal Arts College Faculty." *Academic Questions,* March 29, 2019. www.nas.org/academic-questions/31/2/homogenous_the_political_affiliations_of_elite_liberal_arts_college_faculty.

Langbert, Mitchell, Anthony J. Quain, and Daniel B. Klein. 2016. "Faculty Voter Registration in Economics, History, Journalism, Law, and Psychology." *Econ Journal Watch* 13 (3): 422–451.

Langbert, Mitchell, and Sean Stevens. 2020. "Partisan Registration and Contributions of Faculty in Flagship Colleges." *National Association of Scholars,* January 17, 2020. www.nas.org/blogs/article/partisan-registration-and-contributions-of-faculty-in-flagship-colleges.

Lanier, Jaron. 2019. *Ten Arguments for Deleting Your Social Media Accounts Right Now.* New York: Picador.

Lankov, Andrei. 2013. *The Real North Korea: Life and Politics in the Failed Stalinist Utopia.* New York: Oxford University Press.

Law, Robin. 1985. "Human Sacrifice in Pre-Colonial West Africa." *African Affairs* 84:53–87.

Lee, Elissa. 2002. "Persuasive Storytelling by Hate Groups Online: Examining Its Effects on Adolescents." *American Behavioral Scientist* 45:927–957.

Lee, Jayeon, and Weiai Xu. 2018. "The More Attacks, the More Retweets: Trump's and Clinton's Agenda Setting on Twitter." *Public Relations Review* 44:201–213.

Leese, Daniel. 2011. *Mao Cult: Rhetoric and Ritual in China's Cultural Revolution.* New York: Cambridge University Press.

Lenzer, Anna. 2019. "The Green New Deal's Supporters Hope to Harness Power of Narrative with Federal Writers' Project." *Fast Company,* April 30, 2019. www.fastcompany.com/90341727/the-green-new-deals-supporters-hope-to-harness-power-of-narrative-with-federal-writers-project.

Lerner, Jennifer, Ye Li, Piercarlo Valdesolo, and Karim S. Kassam. 2015. "Emotion and Decision Making." *Annual Review of Psychology* 66:799–823.

Levinson, Ronald. 1953. *In Defense of Plato.* Cambridge, MA: Harvard University Press.

Lewis, C. S. (1959) 2013. *Studies in Words.* New York: HarperCollins.

Lewis-Williams, David. 2002. *The Mind in the Cave.* London: Thames and Hudson.

Liddell, Henry George, and Robert Scott. 1940. *A Greek–English Lexicon.* Oxford, UK: Clarendon Press.

Loxton, Daniel. 2018. "Is the Earth Flat?" *Skeptic*, March 20, 2018. www.skeptic.com/insight/flat-earth-conspiracy-theory/.

Lyons, Leonard. 1956. "The Lyons Den." *Pittsburgh Press,* March 2, 1956, 19.

Mann, Charles. 2011. *1493: Uncovering the New World Columbus Created.* New York: Knopf.

Mar, Raymond A. 2004. "The Neuropsychology of Narrative: Story Comprehension, Story Production and Their Interrelation." *Neuropsychologia* 42:1414–1434.

Mar, Raymond, Keith Oatley, Jacob Hirsh, Jennifer dela Paz, Jordan B. Peterson. 2006. "Bookworms Versus Nerds: Exposure to Fiction Versus Non-Fiction, Divergent Associations with Social Ability, and the Simulation of Fictional Social Worlds." *Journal of Research in Personality* 40:694–712.

Marks, John. 1979. *The Search for the "Manchurian Candidate": The CIA and Mind Control.* New York: Times Books.

Martin, Bradley. 2004. *Under the Loving Care of the Fatherly Leader: North Korea and the Kim Dynasty.* New York: Thomas Dunne Books.

McCabe, Mary Margaret. 2006. "Form and the Platonic Dialogues." In *A Companion to Plato,* edited by Hugh Benson, 39–54. Hoboken, NJ: Blackwell.

McCarthy, Justin. 2020. "U.S. Support for Same-Sex Marriage Matches Record High." *Gallup News,* June 1, 2020. news.gallup.com/poll/311672/support-sex-marriage-matches-record-high.aspx.

McLuhan, Marshall. (1962) 2011. *The Gutenberg Galaxy: The Making of Typographic Man.* Toronto: University of Toronto Press.

McNamee, Roger. 2019. *Zucked: Waking Up to the Facebook Catastrophe.* New York: Penguin.

Mecklin, John. 2017. "Climate Communication: Are Apocalyptic Messages Ever Effective? Interview with Journalist and Science Writer Jon Christensen." *Yale Climate Connections*, August 23, 2017. www.yaleclimateconnections.org/2017/08/climate-communication-do-apocalyptic-messages-work/.

Mekemson, C., and S. Glantz. 2002. "How the Tobacco Industry Built Its Relationship with Hollywood." *Tobacco Control* 11:81–91.

Mercier, Hugo, and Dan Sperber. 2017. *The Enigma of Reason*. Cambridge, MA: Harvard University Press.

Mikkelson, David. 2008. "Did Ernest Hemingway Write a Six-Word Story to Win a Bet?" Snopes.com, October 29, 2008. www.snopes.com/fact-check/hemingway-baby-shoes/.

Mingfu, Liu. 2015. *The China Dream: Great Power Thinking and Strategic Posture in the Post-American Era*. Kindle ed. New York: CN Times.

Miranda, Robbin, William D. Casebeer, Amy M. Hein, Jack W. Judy, Eric P. Krotkov, Tracy L. Laabs, Justin E. Manzo, et al. 2015. "DARPA-Funded Efforts in the Development of Novel Brain-Computer Interface Technologies." *Journal of Neuroscience Methods* 15:52–67.

Mitchell, David. 2001. *Ghostwritten*. New York: Vintage.

Mitchell, Stephen. 1991. *The Gospel According to Jesus: A New Translation and Guide According to Essential Teachings for Believers and Non-Believers*. New York: HarperCollins.

Mithen, Steven. 2009. "Out of the Mind: Material Culture and the Supernatural." In *Becoming Human: Innovation in Prehistoric Material and Spiritual Culture*, edited by Colin Renfrew and Iain Morley, 123–134. Cambridge: Cambridge University Press.

Mittler, Barbara. 2012. *A Continuous Revolution: Making Sense of Cultural Revolution Culture*. Cambridge, MA: Harvard University Press.

Mommsen, Hans. 1999. "German Society and the Resistance Against Hitler." In *The Third Reich: The Essential Readings*, edited by Christian Leitz, 255–273. Hoboken, NJ: Blackwell.

Morin, Olivier, Oleg Sobchuk, and Alberto Acerbi. 2019. "Why People Die in Novels: Testing the Ordeal Simulation Hypothesis." *Palgrave Communications* 5. doi:10.1057/s41599-019-0267-0.

Muchembled, Robert. 2012. *A History of Violence: From the End of the Middle Ages to the Present*. Cambridge: Polity Press.

Mumper, M. J., and R. J. Gerrig. 2017. "Leisure Reading and Social Cognition: A Meta-Analysis." *Psychology of Aesthetics, Creativity, and the Arts* 11:109–120.

Murdock, George. 1949. *Social Structure.* New York: Macmillan.

Murphy, Sheila, Lauren B. Frank, Joyee S. Chatterjee, and Lourdes Baezconde-Garbanati. 2013. "Narrative Versus Nonnarrative: The Role of Identification, Transportation, and Emotion in Reducing Health Disparities." *Journal of Communications* 63:116–137.

Murrar, Sohad, and Markus Brauer. 2018. "Entertainment Education Effectively Reduces Prejudice." *Group Processes and Intergroup Relations* 21:1053–1077.

Myers, B. R. 2010. *The Cleanest Race: How the North Koreans See Themselves—and Why It Matters.* New York: Melville House.

Nabi, Robin, and Melanie Green. 2015. "The Role of a Narrative's Emotional Flow in Promoting Persuasive Outcomes." *Media Psychology* 18:137–162.

Nabi, Robin, Abby Prestin, and Jiyeon So. 2016. "Could Watching TV Be Good for You? Examining How Media Consumption Patterns Relate to Salivary Cortisol." *Health Communication* 31:1345–1355.

Nagel, Thomas. 1979. *Mortal Questions.* Cambridge: Cambridge University Press.

Nails, Debra. 2006. "The Life of Plato of Athens." In *A Companion to Plato,* edited by Hugh Benson, 1–12. Hoboken, NJ: Blackwell.

Nelkin, Dana K. 2019. "Moral Luck." *Stanford Encyclopedia of Philosophy.* Edited by Edward N. Zalta. April 19, 2019. plato.stanford.edu/entries/moral-luck/.

Newman, Nic, with Richard Fletcher, Antonis Kalogeropoulos, David A. L. Levy, and Rasmus Kleis Nielsen. 2018. *Reuters Institute Digital News Report.* Reuters Institute. reutersinstitute.politics.ox.ac.uk/sites/default/files/digital-news-report-2018.pdf?utm_source=digitalnewsreport.org&utm_medium=referral.

Nicholson, Nigel, and Joanne Trautman, eds. 1975–1980. *The Letters of Virginia Woolf.* 6 vols. New York: Harcourt.

Nielsen Global Media. 2020. *Nielsen Total Audience Report: February 2020.* www.nielsen.com/us/en/insights/report/2020/the-nielsen-total-audience-report-february-2020/.

Nietzsche, Friedrich. (1882) 1974. *The Gay Science.* Translated by Walter Kaufmann. New York: Vintage.

Nomura, Ryota, Kojun Hin, Makoto Shimazu, Yingzong Liang, and Takeshi Okada. 2015. "Emotionally Excited Eyeblink-Rate Variability Predicts an Experience of Transportation into the Narrative World." *Frontiers in Psychology* 6. doi:10.3389/fpsyg.2015.00447.

Nomura, Ryota, and Takeshi Okada. 2014. "Spontaneous Synchronization of Eyeblinks During Storytelling Performance." *Bulletin of the Japanese Cognitive Science Society* 21:226–244.

NPR. 2019. "One-Hundred Years Ago This Week, House Passes Bill Advancing 19th Amendment." *Morning Edition*, May 22, 2019. www.npr.org/2019/05/22/725610789/100-years-ago-this-week-house-passes-bill-advancing-19th-amendment.

O'Barr, William. 2013. "'Subliminal' Advertising." *Advertising and Society Review* 13 (4). doi:10.1353/asr.2006.0014.

O'Brien, Edna. 1999. *James Joyce: A Life.* New York: Penguin.

O'Connor, Cailin, and James Weatherall. 2019. *The Misinformation Age: How False Beliefs Spread.* New Haven, CT: Yale University Press.

Office of the Surgeon General. 2017. "Preventing Tobacco Use Among Youths, Surgeon General Fact Sheet." Office of the Surgeon General, 2017. www.hhs.gov/surgeongeneral/reports-and-publications/tobacco/preventing-youth-tobacco-use-factsheet/index.html.

Oksman, Olga. 2016. "Conspiracy Craze: Why 12 Million Americans Believe Alien Lizards Rule Us." *The Guardian*, April 7, 2016. www.theguardian.com/lifeandstyle/2016/apr/07/conspiracy-theory-paranoia-aliens-illuminati-beyonce-vaccines-cliven-bundy-jfk.

O'Malley, Zach, and Natalie Robehmed, eds. 2018. "The Celebrity 100: The World's Highest-Paid Entertainers." *Forbes*, July 16, 2018. www.forbes.com/celebrities/#4caa09845947.

Online Etymology Dictionary. n.d. "Villain." Accessed June 6, 2020. www.etymonline.com/word/villain#etymonline_v_7791.

Ord, Toby. 2020. *The Precipice.* New York: Hachette.

Orlowski, Jeff, director. 2020. *The Social Dilemma* (film). Netflix.

Oschatz, Corinna, and Caroline Marker. 2020. "Long-Term Persuasive Effects in Narrative Communication Research: A Meta-Analysis." *Journal of Communication* 70 (4): 473–496.

Osgood, Kenneth. 2006. *Total Cold War: Eisenhower's Secret Propaganda Battle at Home and Abroad.* Lawrence: University of Kansas Press.

Packard, Vance. 1957. *The Hidden Persuaders.* New York: Ig Publishing.

Paluck, Elizabeth Levy. 2009. "Reducing Intergroup Prejudice and Conflict Using the Media: A Field Experiment in Rwanda." *Journal of Personality and Social Psychology* 96:574–587.

Panero, Maria. E., Deena S. Weisberg, Jessica Black, Thalia R. Goldstein, Jennifer L. Barnes, Hiram Brownell, and Ellen Winner. 2016. "Does Reading a Single Passage of Literary Fiction Really Improve Theory of Mind? An Attempt at Replication." *Journal of Personality and Social Psychology* 111 (5). doi:10.1037/pspa0000067.

Parandowski, Jan. 2015. "Meeting with Joyce." *Studi Irlandesi. A Journal of Irish Studies* 5:135–142.

Parker, Kim. 2019. "The Growing Partisan Divide in Views of Higher Education." *Pew Social Trends*, August 19, 2019. www.pewsocial trends.org/essay/the-growing-partisan-divide-in-views-of-higher -education/.

Perlberg, Steven. 2020. "'There's No Antagonist': News Outlets Mull the Possible End of Their Editorial and Business-Side 'Trump Bump' Bonanza." Digiday, August 10, 2020. digiday.com/media /theres-no-antagonist-news-outlets-mull-the-possible-end-of -their-editorial-and-business-side-trump-bump-bonanza/.

Pettegree, Andrew. 2014. *The Invention of News: How the World Came to Know About Itself.* New Haven, CT: Yale University Press.

Pew Research Center's Forum on Religion and Public Life. 2012. *The Global Religious Landscape.* Pew Research Center, December 18, 2012. www.pewforum.org/2012/12/18/global-religious-landscape-exec/.

Pinker, Steven. 2002. *The Blank Slate.* New York: Viking.

———. 2012. *The Better Angels of Our Nature: Why Violence Has Declined.* New York: Penguin.

———. 2018a. *Enlightenment Now: The Case for Reason, Science, Humanism, and Progress.* New York: Viking.

———. 2018b. "Steven Pinker Recommends Books to Make You an Optimist." *The Guardian*, February 26, 2018. www.theguardian.com /books/2018/feb/26/further-reading-steven-pinker-books-to-make -you-an-optimist.

Piotrow, Phyllis, and Esta de Fossard. 2004. "Entertainment-Education as a Public Health Intervention." In *Entertainment-Education and Social Change: History, Research, and Practice*, edited by Arvind Singhal, M. J. Cody, M. Rogers, and M. Sabido, 39–60. Mahwah, NJ: Lawrence Erlbaum.

Plato. 2016. *The Republic.* Translated by Allan Bloom. New York: Basic Books.

Polti, Daniel. 2020. "Trump Reportedly Considering Launching 2024 Campaign During Biden's Inauguration." *Slate*, November 28, 2020. slate.com/news-and-politics/2020/11/trump-2024-campaign-biden-inauguration.html.

Pomerantsev, Peter. 2019. *This Is Not Propaganda: Adventures in the War Against Reality.* London: Faber and Faber.

Pomeroy, Sarah, Stanley M. Burstein, Walter Donlan, Jennifer Tolbert Roberts, David Tandy, and Georgia Tsouvala. 1999. *Ancient Greece: A Political, Social, and Cultural History.* New York: Oxford University Press.

Poniewozik James. 2019. *Audience of One: Donald Trump, Television, and the Fracturing of America.* New York: Liveright.

Popper, Karl. 1945. *The Open Society and Its Enemies.* London: Routledge.

PornHub. 2019. "The 2019 Year in Review." PornHub Insights, December 11, 2019. www.pornhub.com/insights/2019-year-in-review.

Power, J. Gerard, Sheila Murphy, and Gail Coover. 1996. "Priming Prejudice: How Stereotypes and Counter-Stereotypes Influence Attribution of Responsibility and Credibility Among Ingroups and Outgroups." *Human Communication Research* 23:36–58.

Power, Samantha. 2013. *A Problem from Hell: America and the Age of Genocide.* New York: Basic Books.

Pratkanis, Anthony, and Elliot Aronson. 1991. *Age of Propaganda: The Everyday Use and Abuse of Persuasion.* New York: W. H. Freeman.

Prior, William. 2006. "The Socratic Problem." In *A Companion to Plato,* edited by Hugh Benson, 25–36. Hoboken, NJ: Blackwell.

Pulizzi, Joe. 2012. "The Rise of Storytelling as the New Marketing." *Publishing Research Quarterly* 28:116–123.

Rain, Marina, Elena Cilento, Geoff MacDonald, and Raymond A. Mar. 2017. "Adult Attachment and Transportation into Narrative Worlds." *Personal Relationships* 24:49–74.

Ratajska, Adrianna, Matt I. Brown, and Christopher F. Chabris. 2020. "Attributing Social Meaning to Animated Shapes: A New Experimental Study of Apparent Behavior." *American Journal of Psychology* 133:295–312.

Ratcliffe, R. G. 2018. "Russians Sowed Divisions in Texas Politics, Says U.S. Senate Report." *Texas Monthly*, December 20, 2018.

www.texasmonthly.com/news/russians-sowed-divisions-texas-politics-says-u-s-senate-report/.

Reeves, Byron, and Clifford Nass. 1996. *The Media Equation: How People Treat Computers, Television, and New Media Like Real People and Places.* New York: Cambridge University Press.

Reilly, Conor. 2006. *Selenium in Food and Health.* 2nd ed. New York: Springer.

Reinhart, R. J. 2020. "Fewer in U.S. Continue to See Vaccines as Important." *Gallup News,* January 14, 2020. news.gallup.com/poll/276929/fewer-continue-vaccines-important.aspx.

Rhodes, Robert, ed. 1974. *Winston S. Churchill: His Complete Speeches, 1897–1963.* 8 vols. New York: Chelsea House.

Richardson, Bradford. 2016. "Liberal Professors Outnumber Conservatives Nearly 12 to 1, Study Finds." *Washington Times,* October 6, 2016. www.washingtontimes.com/news/2016/oct/6/liberal-professors-outnumber-conservatives-12-1/.

Ridley, Matt. 2010. *The Rational Optimist: How Prosperity Evolves.* New York: HarperCollins.

Rieff, David. 2016. *In Praise of Forgetting: Historical Memory and Its Ironies.* New Haven, CT: Yale University Press.

Riese, Katrin, Mareike Bayer, Gerhard Lauer, and Annekathrin Schacht. 2014. "In the Eye of the Recipient: Pupillary Responses to Suspense in Literary Classics." *Scientific Study of Literature* 4:211–232.

Roberts, Ian, ed. 2017. *The Oxford Handbook of Universal Grammar.* Oxford: Oxford University Press.

Robinson, David. 2017. "Examining the Arc of 100,000 Stories: A Tidy Analysis." Variance Explained, April 26, 2017. varianceexplained.org/r/tidytext-plots/.

Robson, David. 2017. "How East and West Think in Profoundly Different Ways." BBC Future, January 19, 2017. www.bbc.com/future/article/20170118-how-east-and-west-think-in-profoundly-different-ways.

Rogers, Stuart. 1992–1993. "How a Publicity Blitz Created the Myth of Subliminal Advertising." *Public Relations Quarterly* 37 (Winter): 12–17.

Rosenberg, Alex. 2018. *How History Gets Things Wrong.* Cambridge, MA: MIT Press.

Rosin, Hanna. 2006. "How Soap Operas Can Change the World." *New Yorker,* June 5, 2006. www.newyorker.com/magazine/2006/06/05/life-lessons.

Rowe, Christopher. 2006. "Interpreting Plato." In *A Companion to Plato*, edited by Hugh Benson. Hoboken, NJ: Blackwell.

Saad, Lydia. 2020. "The U.S. Remained Center-Right, Ideologically, in 2019." Gallup News, January 9, 2020. news.gallup.com/poll/275792/remained-center-right-ideologically-2019.aspx.

Sachs, Jonah. 2012. *Winning the Story Wars: Why Those Who Tell (and Live) the Best Stories Will Rule the Future.* Brighton, MA: Harvard Business Review Press.

Samuel, Lawrence. 2010. *Freud on Madison Avenue: Motivation Research and Subliminal Advertising in America.* Philadelphia: University of Pennsylvania.

Santayana, George. 1905. *Reason in Common Sense.* New York: Scribner's.

Sapolsky, Robert. 2018. *Behave: The Biology of Humans at Our Best and Worst.* New York: Penguin.

Scalise Sugiyama, Michelle S. 2003. "Cultural Variation Is Part of Human Nature." *Human Nature* 14:383–396.

Schiappa, Edward, Peter Gregg, and Dean Hewes. 2005. "The Parasocial Contact Hypothesis." *Communications Monographs* 72:92–115.

———. 2006. "Can One TV Show Make a Difference? *Will & Grace* and the Parasocial Contact Hypothesis." *Journal of Homosexuality* 5:15–37.

Schick, Nina. 2020. *Deepfakes: The Coming Infocalypse.* New York: Twelve.

Schmidt, Megan. 2020. "How Reading Fiction Increases Empathy and Encourages Understanding." *Discover Magazine*, August 28, 2020. www.discovermagazine.com/mind/how-reading-fiction-increases-empathy-and-encourages-understanding.

Schulman, Eric. 2016. "Measuring Fame Quantitatively. Who's the Most Famous of Them All? (Part 2)." *Annals of Improbable Research Online*, August 29, 2016. www.improbable.com/wp-content/uploads/2016/08/MEASURING-FAME-part2-2016-09.pdf.

Schulte, Gabriela. 2020. "Many Americans Are Buying into Baseless Theories Around COVID-19, Poll Shows." *The Hill*, August 27, 2020. thehill.com/hilltv/what-americas-thinking/514031-poll-majority-of-americans-buy-into-misinformation-surrounding.

Sestero, Greg, and Tom Bissell. 2013. *The Disaster Artist: My Life Inside The Room, the Greatest Bad Movie Ever Made.* New York: Simon & Schuster.

Sestir, Marc, and Melanie C. Green. 2010. "You Are Who You Watch: Identification and Transportation Effects on Temporary Self-Concept." *Social Influence* 5:272–288.

Shaer, Matthew. 2014. "What Emotion Goes Viral the Fastest?" *Smithsonian Magazine,* April 2014. www.smithsonianmag.com/science-nature/what-emotion-goes-viral-fastest-180950182/.

Shane, Scott, and Mark Mazzetti. 2018. "The Plot to Subvert an Election: Unraveling the Russia Story So Far." *New York Times,* September 20, 2018. www.nytimes.com/interactive/2018/09/20/us/politics/russia-interference-election-trump-clinton.html.

Shapiro, Ben. 2011. *Primetime Propaganda: The True Hollywood Story of How the Left Took Over Our TV.* New York: Broadside Books.

Sharf, Zack. 2019. "Netflix Saved and Collected Every Choice Viewers Made in 'Black Mirror: Bandersnatch.'" Yahoo! Entertainment, February 13, 2019. www.yahoo.com/entertainment/netflix-saved-collected-every-choice-165856905.html.

Sharot, Tali. 2017. *The Influential Mind.* New York: Henry Holt.

Shedlosky-Shoemaker, Randi, Kristi A. Costabile, and Robert M. Arkin. 2014. "Self-Expansion Through Fictional Characters." *Self and Identity* 13:556–578.

Shelley, Percy Bysshe. (1840) 1891. *A Defense of Poetry.* Boston: Ginn & Company.

Shermer, Michael. 2002. *In Darwin's Shadow: The Life and Science of Alfred Russel Wallace.* New York: Oxford University Press.

———. 2015. *The Moral Arc: How Science Makes Us Better People.* New York: Henry Holt.

Shome, Debika, and Sabine Marx. 2009. *The Psychology of Climate Change Communication: A Guide for Scientists, Journalists, Educators, Political Aides, and the Interested Public,* New York: Center for Research on Environmental Decisions.

Shrum, L. 2012. *The Psychology of Entertainment Media: Blurring the Lines Between Entertainment and Persuasion.* New York: Routledge.

Sidney, Philip. (1595) 1890. *The Defense of Poesy.* Cambridge, MA: Harvard University Press.

Sies, Helmut. 2015. "Oxidative Stress: A Concept in Redox Biology and Medicine." *Redox Biology* 4:180–183. doi:10.1016/j.redox.2015.01.002.

Singer, P. W., and Emerson Brooking. 2016. "War Goes Viral: How Social Media Is Being Weaponized." *The Atlantic,* November 2016. www.theatlantic.com/magazine/archive/2016/11/war-goes-viral/501125/.

———. 2018. *LikeWar: The Weaponization of Social Media.* New York: Houghton Mifflin Harcourt.

Singh, Manvir. 2019. "The Sympathetic Plot: Identifying and Explaining a Narrative Universal." Preprint. June 2019. www.researchgate.net/publication/333968807_The_sympathetic_plot_Identifying_and_explaining_a_narrative_universal.

Singhal, Arvind, and Everett Rogers. 2004. "The Status of Entertainment-Education Worldwide." In *Entertainment-Education and Social Change: History, Research, and Practice,* edited by Arvind Singhal, Michael J. Cody, Everett M. Rogers, and Miguel Sabido, 3–20. Mahwah, NJ: Lawrence Erlbaum.

Slater, Michael, Benjamin K. Johnson, Jonathan Cohen, Maria Leonora G. Comello, and David R. Ewoldsen. 2014. "Temporarily Expanding the Boundaries of the Self: Motivations for Entering the Story World and Implications for Narrative Effects." *Journal of Communication* 65:439–455.

Smith, Daniel, Philip Schlaepfer, Katie Major, Mark Dyble, Abigail E. Page, James Thompson, Nikhil Chaudhary, et al. 2017. "Cooperation and the Evolution of Hunter-Gatherer Storytelling." *Nature Communications* 8:1–9.

Soroka, Stuart, Patrick Fournier, and Lilach Nir. 2019. "Cross-National Evidence of a Negativity Bias in Psychophysiological Reactions to News." *Proceedings of the National Academy of Sciences* 116 (38): 18888–18892.

Sorokin, Andrew Ross, Jason Karaian, Michael J. de la Merced, Lauren Hirsch, and Ephrat Livni. 2020. "The Media's Complicated Relationship with Trump: The Industry Has Enjoyed a Boom. Is a Bust Next?" *New York Times,* November 6, 2020. www.nytimes.com/2020/11/06/business/dealbook/media-trump-bump.html.

Soto Laveaga, Gabriela. 2007. "'Let's Become Fewer: Soap Operas, Contraception, and Nationalizing the Mexican Family in an Overpopulated World." *Sexuality Research and Social Policy* 4 (3): 19–33.

Spaulding, Charles B., and Henry A. Turner. 1968. "Political Orientation and Field of Specialization Among College Professors." *Sociology of Education* 41 (3): 247–262.

Stanovich, Keith E. 2015. "Rational and Irrational Thought: The Thinking That IQ Tests Miss: Why Smart People Sometimes Do Dumb Things." *Scientific American,* January 1, 2015. www

.scientificamerican.com/article/rational-and-irrational-thought-the-thinking-that-iq-tests-miss/.

Stark, Rodney. 1996. *The Rise of Christianity: How the Obscure, Marginal Jesus Movement Became the Dominant Religious Force in the Western World in a Few Centuries.* New York: HarperCollins.

Stasavage, David. 2020. *The Decline and Rise of Democracy: A Global History from Antiquity to Today.* Princeton, NJ: Princeton University Press.

Stephens, Greg, Lauren J. Silbert, and Uri Hasson. 2010. "Speaker-Listener Neural Coupling Underlies Successful Communication." *Proceedings of the National Academy of Sciences* 107:14425–14430.

Stephenson, Jill. 2001. "Review Article: Resistance and the Third Reich." *Journal of Contemporary History* 36 (3): 507–516.

Stever, Gayle. 2016. "Evolutionary Theory and Reactions to Mass Media: Understanding Parasocial Attachment." *Psychology of Popular Media Culture* 6 (2): 95–102. doi:10.1037/ppm0000116.

———. 2017. "Parasocial Theory: Concepts and Measures." In *International Encyclopedia of Media Effects,* edited by Patrick Rossler, Cynthia Hoffner, and Liesbet van Zoonen. New York: Wiley. doi:10.1002/9781118783764.wbieme0069.

Stikic, Maja, Robin R. Johnson, Veasna Tan, and Chris Berka. 2014. "EEG-Based Classification of Positive and Negative Affective States." *Brain-Computer Interfaces* 1 (2): 99–112.

Strauss, Leo. 1964. *The City and Man.* Chicago: University of Chicago Press.

Strick, Madelijn, Hanka L. de Bruin, Linde C. de Ruiter, and Wouter Jonkers. 2015. "Striking the Right Chord: Moving Music Increases Psychological Transportation and Behavioral Transportation." *Journal of Experimental Psychology: Applied* 21:57–72.

Strittmatter, Kai. 2020. *We Have Been Harmonized: Life in China's Surveillance State.* New York: Custom House.

Stubberfield, Joseph. 2018. "Contemporary Folklore Reflects Old Psychology." *Evolutionary Studies in Imaginative Culture,* August 27, 2018. esiculture.com/blog/2018/8/27.

Stubberfield, Joseph, Jamshid J. Tehrani, and Emma G. Flynn. 2017. "Chicken Tumours and a Fishy Revenge: Evidence for Emotional Content Bias in the Cumulative Recall of Urban Legends." *Journal of Cognition and Culture* 17:12–26.

Sunstein, Cass. 1999. "The Law of Group Polarization." John M. Olin Program in Law and Economics Working Paper No. 91, 1–39.

———. 2019. *Conformity: The Power of Social Influences.* New York: New York University Press.

Szalay, Jessie. 2016. "What Are Free Radicals?" Live Science, May 27, 2016. www.livescience.com/54901-free-radicals.html.

Tajfel, H., and J. C. Turner. 1986. "The Social Identity Theory of Intergroup Behavior." *Psychology of Intergroup Relations* 5:7–24.

Tal-Or, Nurit, and Yael Papirman. 2007. "The Fundamental Attribution Error in Attributing Fictional Figures' Characteristics to the Actors." *Media Psychology* 9:331–345.

Taylor, A. E. 1926. *Plato: The Man and His Work.* New York: Routledge.

Tarrant, Shira. 2016. *The Pornography Industry.* New York: Oxford University Press.

Thornton, John. 1998. *Africa and Africans in the Making of the Atlantic World, 1400–1800.* Cambridge: Cambridge University Press.

Thu-Huong, Ha. 2018. "These Are the Books Students at the Top US Colleges Are Required to Read." Quartz, January 27, 2018. qz.com/602956/these-are-the-books-students-at-the-top-us-colleges-are-required-to-read/.

Tolstoy, Leo. (1897) 1899. *What Is Art?* Translated by Aylmer Maude. New York: Thomas Y. Crowell.

Tugend, Alina. 2014. "Storytelling Your Way to a Better Job or a Stronger Start-Up." *New York Times,* December 12, 2014, B4.

Van Laer, Tom, Ko de Ruyter, Luca M. Visconti, and Martin Wetzels. 2014. "The Extended Transportation-Imagery Model: A Meta-Analysis of the Antecedents and Consequences of Consumers' Narrative Transportation." *Journal of Consumer Research* 40:797–817.

Vezzali, Loris, Sofia Stathi, Dino Giovannini, Dora Capozza, and Elena Trifiletti. 2015. "The Greatest Magic of Harry Potter: Reducing Prejudice." *Journal of Applied Social Psychology* 45:105–121.

Vonnegut, Kurt. 2018. "Story Shapes According to Kurt Vonnegut: Man in Hole." Filmed September 1, 2018. YouTube video, 1:05. www.youtube.com/watch?v=1w7IueiJHAQ&ab_channel=Ira Portman.

Vosoughi, Soroush, Deb Roy, and Sinan Aral. 2018. "The Spread of True and False News Online." *Science* 359:1146–1151.

Wagner, James Au. 2016. "VR Will Make Life Better—or Just Be an Opiate for the Masses." *Wired*, February 25, 2016. www.wired.com/2016/02/vr-moral-imperative-or-opiate-of-masses/.

Warren, Robert Penn. 1969. *Audubon*. New York: Random House.

Weber, Rene, Ron Tamborini, Hye Eun Lee, and Horst Stipp. 2008. "Soap Opera Exposure and Enjoyment: A Longitudinal Test of Disposition Theory." *Media Psychology* 11:462–487.

Weinberger, Sharon. 2014. "Building the Pentagon's 'Like Me' Weapon." BBC Future, November 18, 2014. www.bbc.com/future/article/20120501-building-the-like-me-weapon.

Weldon, Laura Grace. 2011. "Fighting 'Mean World Syndrome.'" *Wired*, January 27, 2011. www.wired.com/2011/01/fighting-mean-world-syndrome/.

Whitley, David S. 2009. *Cave Paintings and the Human Spirit: The Origin of Creativity and Belief*. Amherst, NY: Prometheus.

Wiessner, Polly. 2014. "Embers of Society: Firelight Talk Among the Ju/'hoansi Bushmen." *Proceedings of the National Academy of Sciences* 111:14027–14035.

Wikipedia. 2021. "List of Highest Grossing Films." Last modified March 11, 2021, accessed January 26, 2021. en.wikipedia.org/wiki/List_of_highest-grossing_films#Highest-grossing_franchises_and_film_series.

Williams, Bernard. 1981. *Moral Luck*. New York: Cambridge University Press.

Wilson, David. 2003. *Darwin's Cathedral: Evolution, Religion, and the Nature of Society*. Chicago: University of Chicago Press.

Wilson, Timothy, David A. Reinhard, Erin C. Westgate, Daniel T. Gilbert, Nicole Ellerbeck, Cheryl Hahn, Casey L. Brown, and Adi Shaked. 2014. "Just Think: The Challenges of the Disengaged Mind." *Science Magazine* 345:75–77.

Wise, Jeff. 2009. *Extreme Fear: The Science of Your Mind in Danger*. New York: Palgrave Macmillan.

World Health Organization. *WHO Report on the Global Tobacco Epidemic, 2008*. World Health Organization, 2008. www.who.int/tobacco/mpower/2008/en/.

Wright, Richard. 2010. *The Evolution of God*. New York: Back Bay Books.

Wu, Tim. 2016. *The Attention Merchants: The Epic Scramble to Get Inside Our Heads*. New York: Knopf.

Wylie, Christopher. 2019. *Mindf*ck: Cambridge Analytica and the Plot to Break America.* New York: Random House.

Zak, Paul J. 2013. *The Moral Molecule: How Trust Works.* New York: Plume.

———. 2015. "Why Inspiring Stories Make Us React: The Neuroscience of Narrative." *Cerebrum* 2015:2.

Zebregs, S., Bas van den Putte, Peter Neijens, and Anneke de Graaf. 2015. "The Differential Impact of Statistical and Narrative Evidence on Beliefs, Attitude, and Intention: A Meta-Analysis." *Health Communication* 30 (3): 282–289.

Zillmann, Dolf. 2000. "Humor and Comedy." In *Media Entertainment: The Psychology of Its Appeal*, edited by D. Zillmann and P. Vorderer, 37–57. Mahwah, NJ: Lawrence Erlbaum.

Zillmann, Dolf, and Joanne R. Cantor. 1977. "Affective Responses to the Emotions of a Protagonist." *Journal of Experimental Social Psychology* 13 (2): 155–165.

Zimmerman, Jess. 2017. "It's Time to Give Up on Facts." *Slate,* February 8, 2017. slate.com/technology/2017/02/counter-lies-with-emotions-not-facts.html.

Zuboff, Shoshana. 2020. *The Age of Surveillance Capitalism: The Fight for a Human Future at the New Frontier of Power.* New York: PublicAffairs.

NOTES

Introduction

1. Brinthaupt 2019; see also Geurts 2018; Kross 2021.
2. Gould 1994, 282.
3. Colapinto 2021.
4. Green and Clark 2012.
5. Oksman 2016.
6. Baird 1974.
7. Ahren 2020; Baird 1974.
8. I adapted this term from E. M. Forster's *Aspects of the Novel* (1927). *Homo fictus* is Forster's term for fiction characters in general, which he amusingly differentiates from actual *Homo sapiens.*
9. Davies, Cillard, Friguet et al 2017.
10. On oxidative stress and "the oxygen paradox," see Davies 2016; Davies and Ursini 1995; Sies 2015; Szalay 2016. On certain nutrients as essential poisons, see Reilly 2006.
11. On the pivotal role of storytelling in human evolution, see Boyd 2009; Boyd, Carroll, and Gottschall 2010; Gottschall 2012; Harari 2015.

Chapter 1: "The Storyteller Rules the World"

1. Nielsen Global Media 2020.
2. Wise 2009.

3. For storytelling's effects on the brain, see Krendl et al. 2006; Morteza et al. 2017; Stephens, Silbert, and Hasson 2010. For lack of peripheral awareness, see Bezdek and Gerrig 2017; Cohen, Shavalian, and Rube 2015. For pupillary and blink responses, see Kang and Wheatley 2017; Nomura and Okada 2014; Nomura et al. 2015; Riese et al. 2014. For endorphin spike and increased pain tolerance, see Dunbar et al. 2016. For correlations of pleasure with measurable physiological responses, see Andersen et al. 2020.

4. Burke and Farbman 1947. For general information on Khoisan storytelling, see Wiessner 2014.

5. For proposed solutions to the evolutionary riddle of storytelling, see Boyd 2009; contributors to Boyd, Carroll, and Gottschall 2010; Carroll et al. 2012; Dissanayake 1990, 1995; Dutton 2009; Gottschall 2012; Pinker 2002.

6. Damasio 2010, 293; see also Asma 2017, 152.

7. Smith et al. 2017.

8. O'Malley and Robehmed 2018.

9. Flavel et al. 1990.

10. Bezdek, Foy, and Gerrig 2013; Dibble and Rosaen 2011; Giles 2002; Hall 2019; Rain et al. 2017; Schiappa, Gregg, and Hewes 2005, 2006; Singh 2019; Stever 2016, 2017.

11. Tal-Or and Papirman 2007.

12. Cantor 2009.

13. For overviews of narrative transportation, see Bezdek and Gerrig 2017; Gerrig 1993; Green and Brock 2000; Green and Dill 2013; Strick et al. 2015; van Laer et al. 2014.

14. Green and Brock 2000.

15. van Laer et al. 2014.

16. Shapiro 2011, xx.

17. Shapiro 2011, xii.

18. See Dreier 2017; Drum 2018; Ellis and Stimson 2012. For the liberalizing trend measured over centuries, see Pinker 2018a.

19. McCarthy 2020.

20. Ellithorpe and Brookes 2018.

21. Shedlosky-Shoemaker, Costabile, and Arkin 2014.

22. Dibble and Rosaen 2011; Schiappa, Gregg, and Hewes 2005, 2006.

23. For diminishing prejudice when whites watch shows with likable black characters, see Power, Murphy, and Coover 1996. For similar ef-

fect with television characters who are Muslim or have disabilities, see Murrar and Brauer 2018. For studies of the prosocial impact of popular fiction in international settings, see Paluck 2009; Piotrow and Fossard 2004; Rosin 2006; Singhal and Rogers 2004; Soto Laveaga 2007

24. Murrar and Brauer 2018. For the limited efficacy of diversity training, see Dobbin and Kale 2013; Chang et al. 2019.

25. Coleridge 1817, chap. 14.

26. See Wilson et al. 2014.

27. On the scientific plausibility of escapist theories of storytelling, see Slater et al. 2014.

28. See Copeland 2017; Killingsworth and Gilbert 2010; Wilson et al. 2014; Kross 2021.

29. Corballis 2015; Kross 2021.

30. Nicholson and Trautman 1975–1980, 5:319.

31. Barraza et al. 2015; Dunbar et al. 2016; Nabi, Prestin, and So 2016; Zak 2013, 2015.

32. For the metaphor of story's active ingredients, see Green 2008, 48.

Chapter 2: The Dark Arts of Storytelling

1. My views on Plato's *Republic* have been informed by outstanding works of scholarship, including Arieti 1991; Benson 2006; contributors to Blondell 2002; Bloom 1968; Hamilton 1961; Havelock 1963; Howland 1993; Janaway 1995, 2006; Kirsch 1968; Levinson 1953; Popper 1945; Pomeroy et al. 1999; Strauss 1964; Taylor 1926.

2. Van Laer et al. 2014, 798.

3. For a sample of an ongoing flood of business books on storytelling, see Gallo 2016; Godin 2012; Guber 2011; Sachs 2012.

4. Singer and Brooking 2016, 2018.

5. Halper 2013.

6. For information on James Vicary, see Crandall 2006; Crispin Miller 2007; Pratkanis and Aronson 1991; Rogers 1992–1993; Samuel 2010.

7. For mid-twentieth-century fears of mind control, see Holmes 2017; Jacobsen 2015; Kinzer 2020; Marks 1979.

8. Quoted in Samuel 2010, 95.

9. Quoted in Samuel 2010, 95.

10. Haberstroh 1994; O'Barr 2013; Rogers 1992–1993; Samuel 2010.

11. O'Barr 2013.

12. Tugend 2014; see also Pulizzi 2012.

13. Godin 2012.

14. Guber 2011.

15. Krendl et al. 2006; Morteza et al. 2017; Stephens et al. 2010. For hormonal harmony, see Zak 2013, 2015; Dunbar et al. 2016. For further evidence of physiological synchronization, see Bracken et al. 2014.

16. Bower and Clark 1969; Dahlstrom 2014; Graesser et al. 1980; Haidt 2012a, 281; Kahneman 2011, 29. In the World Memory Championships, people compete to see who can memorize the most the fastest. In his book *Moonwalking with Einstein* (2012), Joshua Foer shows that success comes down more to storytelling technique than to the innate capacity of memory. By working information into stories, competitors accomplish seemingly impossible feats of recall—like memorizing huge lists of random words or numbers.

17. See Chapter 3.

18. Quoted in Lyons 1956.

19. Damasio 2005; Lerner et al. 2015.

20. Bail et al. 2018; Kolbert 2017.

21. See meta-analyses by Braddock and Dillard 2016; Oschatz and Marker 2020; van Laer et al. 2014. See also Brechman and Purvis 2015; De Graaf and Histinx 2011; Green and Clark 2012; Green and Dill 2013; Murphy et al. 2013; Nabi and Green 2015; Shrum 2012; Strick et al. 2015. But some studies show no advantage in persuasiveness of story-based communication: Allen and Preiss 1997; Ecker, Butler, and Hamby 2020; Zebregs et al. 2015.

22. Van Laer et al. 2014.

23. Mikkelson 2008.

24. Dahlstrom 2014; see also Lee 2002.

25. Hamby and Brinberg 2016; Hamby, Brinberg, and Daniloski 2017.

26. Didion 1976, 270.

27. Gardner 1978, 39.

28. Gardner 1983, 87.

29. Sharf 2019.

30. An able overview of studies of this type (and many others) is found in Dill-Shackleford and Vinney 2020.

31. Del Giudice, Booth, and Irwing 2012; Jarrett 2016.

32. See Brechman and Purvis 2015; Chen 2015. On sex differences in transportability, see van Laer et al. 2014.

33. Jacobsen 2015, 7.

34. Cha 2015; Defense Advanced Research Projects Agency. n.d.; Weinberger 2014.

35. Barraza et al. 2015. For similar studies, see Correa et al. 2015; Stikic et al. 2014. For an overview of DARPA efforts to create brain–computer interfaces, see Miranda et al. 2015. For DARPA-funded attempts to use transcranial magnetic simulation to alter narrative processing, see Corman et al. 2013.

36. The People's Republic of China is in the process of illustrating worst-case scenarios for how such data could be swept up and used toward repressive ends: Anderson 2020; Strittmatter 2020. On the use of new technology for mind reading and mind control, see Fields 2020.

37. Bentham 1791.

38. Lanier 2019, 8. For information on the online architecture of big data and behavioral control also, see Orlowski 2020; Wu 2016; Wylie 2019; Zuboff 2020.

39. Pomerantsev 2019; Shane and Mazzetti 2018.

40. Shane and Mazzetti 2018.

41. Pomerantsev 2019; Ratcliffe 2018; Shane and Mazzetti 2018; Wylie 2019.

42. Thu-Huong 2018. For a constantly updated indicator of the most frequently assigned texts in college courses, see the Open Syllabus Project at opensyllabus.org/.

43. Sidney (1595) 1890, 41.

Chapter 3: The Great War for Storyland

1. Grube 1927.

2. On the control of storytelling (and other art forms) in totalitarian societies, see Arendt (1948) 1994. For Maoist China, see Chiu and Shengtian 2008; Leese 2011; Mittler 2012. For North Korea, see Lankov 2013; Martin 2004; Myers 2010. For the Soviet Union, see Brandenberg 2011; Kenez 1974; Osgood 2006. Similar dynamics are in play in the mainly right-wing authoritarian movements that have swept much of the world in recent years but which are even more focused on narrative dominance rather than jackbooted repression (Guriev and Treisman 2019).

3. Tolstoy (1897) 1899, chap. 15.

4. Berger 2012; AVAAZ 2020; Heath, Bell, and Sternberg 2001; Stubbersfield 2018; Vosoughi, Roy, and Aral 2018.

5. Nabi and Green 2015, 151.

6. Stubberfield 2018; see also Heath, Bell, and Sternberg 2001; Stubberfield et al. 2017.

7. Brady et al. 2017; see also Lee and Xu 2018.

8. Berger 2013; Berger and Milkman 2012; Brady et al. 2017; Hamby and Brinberg 2016; Jones, Libert, and Tynski 2016; Shaer 2014; Sharot 2017; Vosoughi, Roy, and Aral 2018.

9. My account of the early Jesus movement is based largely on several books by Bart Ehrman (2007, 2014, and especially 2018). Other sources consulted include Mitchell 1991 and Stark 1996.

10. The Pew Research Center for Religion and Public Life found that 31.5 percent of the world's population is affiliated with the Christian religion. The religion with the second highest number of followers is Islam, with 23.2 percent of the world's population (see Pew Research Center's Forum on Religion and Public Life 2012).

11. Pew Research Center's Forum on Religion and Public Life 2012.

12. Ehrman 2018, 119.

13. Ehrman 2018; see also Stark 1996.

14. Ehrman 2018, 153. In his book *Heaven and Hell* (2020), Ehrman stresses that consistent visions of the afterlife can't be derived from the Bible and only evolved gradually after the time of Christ.

15. Ehrenreich 2021; Enten 2017; Ghose 2016; Henley and McIntyre 2020; O'Connor and Weatherall 2019; Schulte 2020.

16. Vosoughi, Roy, and Aral 2018.

17. Quoted in Havelock 1963, 4.

18. Mar 2004, 1414.

19. Campbell 1949.

20. My primary on the flat Earth movement is Garwood 2007. Other sources include Burdick 2018; Loxton 2018; Shermer 2002. The flat Earth movement is dated to the nineteenth century because, contrary to what you may have heard about Columbus, there has never been a time in the last couple thousand years when educated people thought the Earth was flat.

21. Branch and Foster 2018.

22. Hambrick and Burgoyne 2016; Sharot 2017, 22–23; Stanovich 2015.

23. See Garwood 2007.

24. From the computer scientist Guillaume Chaslot, quoted in Orlowski 2020.

25. Franks, Bangerter, and Bauer 2013. On the QAnon movement as a quasi-religious movement, see LaFrance 2020.

26. Lenzer 2019; Mecklin 2017; Shome and Marx 2009.

27. Baumeister et al. 2001; Fessler, Pisor, and Navarrete 2014; see also Soroka, Fournier, and Nir 2019.

Chapter 4: The Universal Grammar

1. This discussion of *Finnegans Wake* draws on material from a previously published article (Gottschall 2013).

2. My sources on James Joyce and *Finnegans Wake* include Bowker 2011; Hayman 1990; O'Brien 1999.

3. From a letter from Joyce to Harriet Weaver, quoted in Hayman 1990, 36.

4. O'Brien 1999, 146.

5. Bloom 1994, 422.

6. Parandowski 2015, 141.

7. O'Brien 1999.

8. See Chomsky 1965. For an overview of modern thinking about the universal grammar, see Roberts 2017. For critique, see Dabrowska 2015; Dor 2015.

9. It's been argued that cultural particularity makes certain stories, such as *Hamlet*, all but indecipherable to people in different cultures (Bohannon 1966). But literary scholar Michelle Scalise Sugiyama (2003) has dismantled this argument and shown that even people from very different cultures have no trouble following stories such as *Hamlet* once cultural differences are explained.

10. Nietzsche (1882) 1974, 74.

11. For more than a half-century, the overwhelming rule in critical theory has been to systematically deny that there are inborn regularities in human nature driving regularities in stories that can be found universally across human populations. See Gottschall 2008; Hogan 2003.

12. As Gail Dines writes in *Pornland* (2011), finding reliable data on pornography consumption is difficult. For informed speculation on the scale of porn consumption, see *The Pornography Industry* by Shira Tarrant (2016). However, all sources agree that pornography consumption is almost unfathomably vast wherever people have internet access. A strong, if incomplete, indicator of the scale of porn consumption is provided in yearly statistics aggregated by PornHub.

In 2019 alone, *169 years'* worth of porn was uploaded onto PornHub, with people consuming more than five billion hours of porn in a grand total of forty-five billion visits. PornHub is the world's most dominant porn site. But these figures still leave out the traffic that goes to millions of other porn sites, many of which also have very high traffic. See PornHub 2019.

13. Robinson 2017.

14. For an explanation of the negativity bias in storytelling, see Gottschall 2012.

15. *Hamlet* Act 3, scene 2.

16. Morin, Sobchuk, and Acerbi 2019.

17. Santayana 1905, 284.

18. Muchembled 2012, 263.

19. Muchembled 2012; Pettegree 2014.

20. Pinker 2012, 2018a. For similarly optimistic arguments, see Ridley 2010; Shermer 2015; and the list of recommendations in Pinker 2018b.

21. Klein 2020.

22. Ord 2020.

23. Grabe 2012; see also Soroka, Fournier, and Nir 2019.

24. Pinker 2018a, 35.

25. Plato 2016, 359.

26. Appel 2008; Gerbner et al. 2006; Weldon 2011.

27. On the connection between mean worldviews and news consumption, see Appel 2008; Dahlstrom 2014.

28. James 2018.

29. Kahneman 2011, 207.

30. Vonnegut 2018.

31. For a classic statement on the deep morality of fiction, see Gardner 1978; see also Carroll et al. 2012.

32. Singh 2019.

33. Zillmann 2000. On the same pattern in stories told to children, see Zillmann and Cantor 1977.

34. Weber et al. 2008.

35. Boehm 2001.

36. Kjeldgaard-Christiansen 2016, 109, 116.

37. For earlier arguments that art behavior evolved partly to promote social cohesion, see Darwin 1871; Dissanayake 1990, 1995.

38. Dunbar et al. 2016.
39. Smith et al. 2017.
40. See Cunliffe 1963, 204; Liddell and Scott 1940.
41. Zillmann 2000, 38.
42. Frye (1957) 2020, 47.
43. Rosenberg 2018, 244.

Chapter 5: Things Fall Apart

1. For historical overviews of the Rwandan genocide, see Gourevitch 1998; Power 2013.
2. Paluck 2009.
3. Argo, Zhu, and Dahl 2008; Djikic and Oatley 2013; Kuzmičová et al. 2017; Mar et al. 2006; Mumper and Gerrig 2017; Schmidt 2020. On the powerful form of story-based empathy known as identification, see Hall 2019; Hoeken, Kolthoff, and Sanders 2016; Hoeken and Sinkeldam 2014; Nabi and Green 2015; Sestir and Green 2010.
4. Vezzali et al. 2015.
5. Djikic and Oatley 2013; Johnson et al. 2013; Kidd and Castano 2013; Mar et al. 2006. For an overview of research, see Mumper and Gerrig 2017. For an overview of studies of the moral effects of storytelling, see Hakemulder 2000. For a failure to replicate empathy effects, see Panero et al. 2016.
6. See Gardner 1978, 147.
7. For role of Hutu Power propaganda in Rwandan genocide, see Gourevitch 1998.
8. Bloom 2016, 31. For further treatment of the downsides of empathy, see Brinthaupt 2019; Sapolsky 2018.
9. Bloom 2016. For a broader description of all the factors driving suicide bombers, see Atran 2003, 2006.
10. Burroway 2003, 32.
11. Barnes and Bloom 2014.
12. Breithaupt 2019, 17.
13. See Bietti, Tilston, and Bangerter 2018; Tajfel and Turner 1986.
14. Quoted in Rieff 2016, 138.
15. Rosenberg 2018, 29.
16. Described in Rieff 2016.
17. Santayana 1905, 284.

18. Rieff 2016, 87.

19. A phrase from the Irish writer Hubert Butler, quoted in Rieff 2016, 39.

20. Rieff 2016, 64.

21. Rosenberg 2018, 5.

22. Rosenberg 2018, 246.

23. Plato's *Republic*, Book III.

24. Plato's *Republic*, Book X.

25. Plato's *Republic*, Book III.

26. Baldwin 1992, 101–102.

27. Chua 2007, xxv.

28. For an overview of philosophical thinking around moral luck, see Hartman 2019; Nelkin 2019. For seminal philosophical papers, see also Nagel 1979 and Williams 1981.

29. This example was inspired by similar musings about the Nazis in Nagel 1979.

30. Mommsen 1999; Stephenson 2001.

31. Law 1985.

Chapter 6: The End of Reality

1. Heider and Simmel 1944. For research confirming Heider and Simmel's conclusions, see Klin 2000; Ratajska, Brown, and Chabris 2020.

2. Heider 1983, 148.

3. Kahneman 2011.

4. For overview of research, see Sapolsky 2018, 455; also Haidt 2012a.

5. Harris 2012, 9.

6. See Kahneman 2011; Sapolsky 2018.

7. Harris 2012, 45.

8. Tolstoy (1897) 1899, 43.

9. McLuhan (1962) 2011.

10. Poniewozik 2019.

11. Lanier 2019, 20.

12. Poniewozik 2019.

13. Pinker 2018a; Ridley 2010.

14. Kessler 2021.

15. Schulman 2016.

16. Newman et al. 2018; Perlberg 2020; Sorokin et al. 2020. For the "Trump Bump" in book publishing, see Alter 2020.

17. Kahn 2020; Kim 2020.

18. Polti 2020.

19. Cillizza 2014.

20. Langbert, Quain, and Klein 2016.

21. Sunstein 1999, 2019.

22. Saad 2020.

23. Jaschik 2016, 2017; Langbert 2019; Langbert, Quain, and Klein 2016; Langbert and Stevens 2020.

24. Langbert 2019.

25. Parker 2019; see also Jaschik 2018; Jones 2018. Incidentally, similar polls have found that nearly 90 percent of Republicans reported trusting the mass media either "not very much" or "not at all" (Brenan 2020).

26. Kashmir and White 2020.

27. Plato 2016, 290.

28. Bazelon 2020; Schick 2020.

29. Bazelon 2020; Pomerantsev 2019. For President Barack Obama's view of post-truth as the "single biggest threat to our democracy," see Goldberg 2020.

30. Rhodes 1974, 7:7566.

31. See Boehm 2001; Stasavage 2020.

32. The process whereby smaller-scale societies with democratic traditions gave way to larger-scale autocracies is told in detail in Stasavage 2020.

33. Vosoughi, Roy, and Aral 2018.

34. AVAAZ 2020.

35. For regulatory suggestions, see Zuboff 2020.

36. Bietti, Tilston, and Bangerter 2018.

37. See also Anderson 2020; *Frontline* 2020; Strittmatter 2020.

38. Mingfu 2015, loc. 457.

39. Robson 2017.

Conclusion: A Call to Adventure

1. This discussion of the prehistoric sculpture in the Tuc D'Audoubert caves draws on material from a previously published article (Gottschall 2016).

2. For sources on the clay bison of the Tuc D'Audoubert caves, see Begouen et al. 2009; Breuil 1979; Brodrick 1963; Lewis-Williams 2002; Whitley 2009.

3. In a recent article in *Nature*, the archaeologist Maxime Aubert and her colleagues (2019) argue, "Humans seem to have an adaptive predisposition for inventing, telling and consuming stories. Prehistoric cave art provides the most direct insight that we have into the earliest storytelling, in the form of narrative compositions or 'scenes'" (442). For similar arguments on prehistoric art, see Azéma and Rivère 2012; Mithen 2009.

4. Popper 1945.

5. See Shelley (1840) 1891; Sidney (1595) 1890.

6. Homer's *Iliad*, Book 14, lines 86–87. My translation.

7. Pomeroy et al. 1999.

INDEX

© Jared White

Jonathan Gottschall is a distinguished research fellow in the English department at Washington & Jefferson College and the author of *The Storytelling Animal*, a *New York Times* Editors' Choice, and *The Professor in the Cage*, one of the *Boston Globe*'s Best Books of the Year. He lives in Washington, Pennsylvania.